An Illustrated Technique for Dilation and Evacuation

Janice Lee, M.D.

Copyright © 2019 Janice Lee, M.D.

Interior Image Credit: Janice Lee, M.D.

The author of this publication has found the techniques explained and illustrated herein to be safe and effective in her hands and believes the information presented to be consistent with generally accepted practices. The author and publisher are not responsible for any errors or omissions or for any consequences arising from the application of the information contained herein and make no warranties, expressed or implied, with regard to the content of this publication. The application of the information presented in this book remains the professional responsibility of the practitioner. Neither the author nor the publisher shall be liable for any damages arising herefrom.

All rights reserved. No part of this book may be reproduced, stored, or transmitted by any means—whether auditory, graphic, mechanical, or electronic—without written permission of the author, except in the case of brief excerpts used in critical articles and reviews. Unauthorized reproduction of any part of this work is illegal and is punishable by law.

ISBN: 978-1-6847-0068-4 (sc)
ISBN: 978-1-6847-0069-1 (e)

Because of the dynamic nature of the Internet, any web addresses or links contained in this book may have changed since publication and may no longer be valid. The views expressed in this work are solely those of the author and do not necessarily reflect the views of the publisher, and the publisher hereby disclaims any responsibility for them.

Any people depicted in stock imagery provided by Getty Images are models, and
such images are being used for illustrative purposes only.
Certain stock imagery © Getty Images.

Lulu Publishing Services rev. date: 05/09/2019

Preface

Few other subjects create the degree of political controversy, moral and legal debate, and personal anguish that abortion does. Second trimester abortion may be performed for emergency indications or for a myriad of medical, social or personal reasons. I wish to remind all readers that the provision of technical information regarding the performance of abortion is not, per se, a moral issue. It is my purpose in this treatise to document and share my technique for Dilation and Evacuation, the usual surgical technique for performing second trimester surgical abortion. Although textbooks such as *Management of Unintended and Abnormal Pregnancy* (1) contain much helpful information, detailed descriptions and illustrations of D&E techniques are lacking. It is my hope that this information will improve the care of women undergoing second trimester abortion for any reason.

Like many other abortion providers of my generation, I was self-trained. These are techniques I developed over an 18-year period while practicing at a large abortion clinic, performing, I estimate, approximately 10,000 D&E procedures. I have personally utilized the techniques illustrated for gestations between 13 and 21 wks., 6 days. This composition was developed specifically for the Ob-Gyn Residents I was training. I believe it will be helpful to other physicians who are learning or teaching D&E, or wish to expand their current knowledge of techniques. It is not my intention to replace one-on-one instruction in D&E by an experienced provider or to negate other techniques, but rather to illustrate and share techniques which I have found successful over the years.

In order to accurately illustrate maneuvers, I have utilized photographs of models. The resulting illustrations are significantly easier to understand than the usual drawings.

Table of Contents

Preface iii

Day 1 1
   Patient preparation 1
   Laminaria insertion 1

Day 2 4
   Pre-op preparation 4
   Instruments for D&E 4
   Operative preparation for extraction 6
   Ultrasound placement and suction 7
   Removing the speculum while leaving the cervical tenaculum in place 8
   Abortion forceps 10
   Models for demonstration of maneuvers 13
   Operative plan 14
   Description and illustration of basic extraction techniques 15
      Approach from lateral 16
      Approach from anterior 20
      Approach from below 22
      Management of "slip" 23
   Management of transverse lie 24
      Conversion to breech 24
      Approach from patient's right 26
      Conversion to shoulder presentation 27
   Extraction with "turn-and-pull" technique 28
   Extraction of remaining parts 30
   Extraction techniques for "entrapped" calvarium 31
      Cervical entrapment 31
      Cornual entrapment 33
   Removal of placenta 36
   Completion of procedure 36
   Management of post-op bleeding 36
   Transfer to post-op 37

References: 38

DAY 1

Patient preparation

Suffice it to say that patients should have an adequate medical and surgical history, which may be taken by a nurse or trained counselor, but should be reviewed by the surgeon. The patient should have completed routine blood work and obtained an ultrasound documenting gestational age, plurality, and placental localization. Always review the ultrasound yourself and repeat anything that appears questionable. Adequate counseling regarding the patient's decision is extremely important since D&E is oftentimes a two-step procedure. In most clinics, counseling is given, consent is obtained and forms signed by a trained counselor or nurse. Verify that forms have been signed. When meeting the patient, introduce yourself, review any medical issues, and ask the patient if she has any questions. A standard "time out" should be performed by the surgeon if not by personnel.

It is important to answer all questions calmly and truthfully in a reassuring manner. The last question to ask the patient is if she is certain of her decision, explaining that once laminaria have been inserted or drugs have been given to induce cervical ripening, she is committing herself to completing the procedure. Any patient who expresses ambivalence or hesitation should receive a description of the risks of continuing a pregnancy after laminaria have been inserted then removed, or cervical ripening agents have been administered. Any patient who does not clearly state her decision to proceed should receive further counseling. Additional risks may be described at the discretion of the surgeon.

Laminaria insertion

Instruments needed for laminaria insertion are basic. See photo below. Since I generally do not place a tenaculum on the cervix, I usually utilize a medium Graves speculum. Numerous other models and sizes are available (pages 4-5). Also required: a prep bowl with solution, sterile lubricating jelly, sponges for prepping and insertion, and Bozeman forceps for grasping the laminaria, as illustrated below.

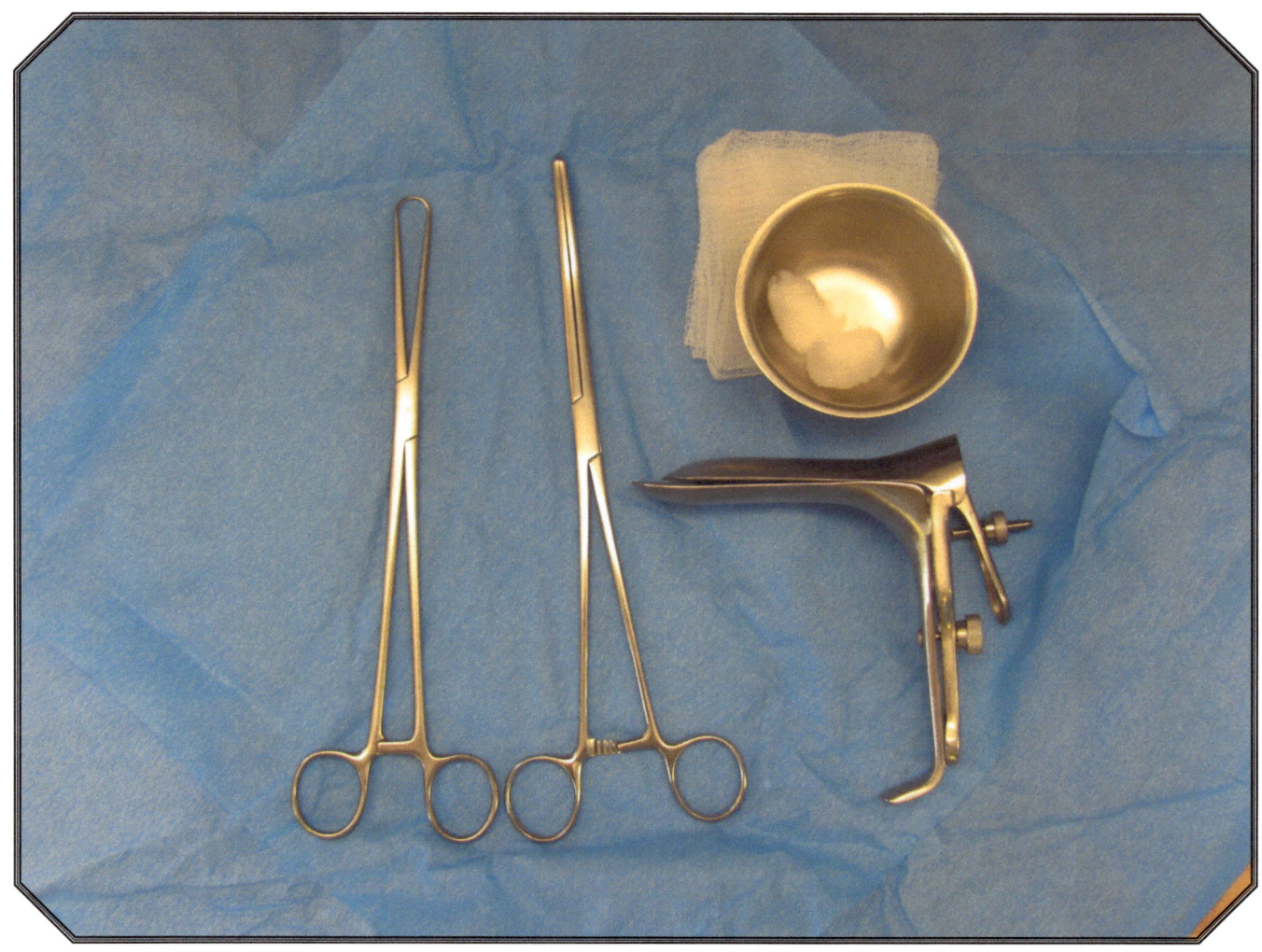

A single-toothed tenaculum, syringe with 15 cc lidocaine 1%, and spinal needle should be available in the event that paracervical block is necessary.

After positioning the patient, proceed to a bimanual exam. Confirm that uterine size and ultrasound dates coincide. The location, consistency, and curvature of the cervix are of particular importance since I prefer not to put a tenaculum on the cervix for insertion of laminaria. For insertion, generally utilize a standard Graves speculum. Place the speculum, tilting the cervix, so that the external cervical os is visible. Prep the cervix with betadine, then wipe dry. A Bozeman forceps should be utilized to grasp the laminaria. I utilize #4 medium size laminaria; two for 16 wks., three for 17 wks., four for 18 wks., and, if possible, five or six for 19-21 wks.

Sterile lubricating jelly should be on the tray. Utilize the Bozeman forceps to grasp the distal half of the laminaria lengthwise initially. Lubricate the tip. Place the distal tip of the laminaria into the canal; advance following the curvature of the canal, by angling the laminaria anterior or posterior if resistance is met, until just the tip and the string are visible. Place the additional laminaria at the side, anterior or posterior, utilizing the first laminaria as a guide. It is common for the laminaria to keep coming out during insertion. Keep repositioning them. If necessary, pause and hold the laminaria in place with the Bozeman until it stabilizes, asking the patient to breathe deeply through an open mouth to reduce abdominal pressure. All strings and tips should be visible at the external os at the end of the insertion. If one "disappears", pull on the string until the tip is seen before proceeding with inserting the next laminaria.

If it is necessary to straighten the cervix, give 2 cc of 1% lidocaine at the tenaculum site before applying the tenaculum. Place the tenaculum on the posterior lip of a retroverted uterus. Full paracervical block should be reserved for difficult cases, or those in which it is necessary to dilate the cervix. Do not force laminaria. This can lead to perforation or difficult removal due to "dumbelling". If few laminaria can be placed in an advanced gestational age situation, placing additional dilapan or giving additional doses of misoprostol the next day may be necessary prior to evacuation. Administering mifepristone on the day of laminaria insertion is another option. To hold laminaria in place, I place a sponge against the cervix with the Bozeman, remove the tenaculum if present, remove the speculum to the handle of the Bozeman, then release the sponge, leaving it in the vagina, and remove the Bozeman and remainder of the speculum at the same time. Patients should be reassured that laminaria may fall out but that this does not constitute a problem. Patients should be given standard written instructions regarding returning on day 2.

DAY 2

Pre-op preparation

For a number of reasons, I have settled on 600 mcg of buccal misoprostol 1 hr. prior to procedure as routine preparation for D&E at 16 wks. or above. Although studies have shown that cervical dilation is not improved by utilizing 600 mcg in comparison to 400 mcg, I do find that the uterotonic activity of 600 mcg is helpful, especially since we have been unable to utilize vasopressin due to exorbitant cost. If the cervix was very soft and patulous at the exam prior to laminaria insertion, shorten the time to 40 min. Patients should swallow any misoprostol not dissolved prior to being brought to the O.R. I have not found intervals longer than 1 hr. to improve results.

Instruments for D&E

Below are various views of the specula I have routinely utilized.

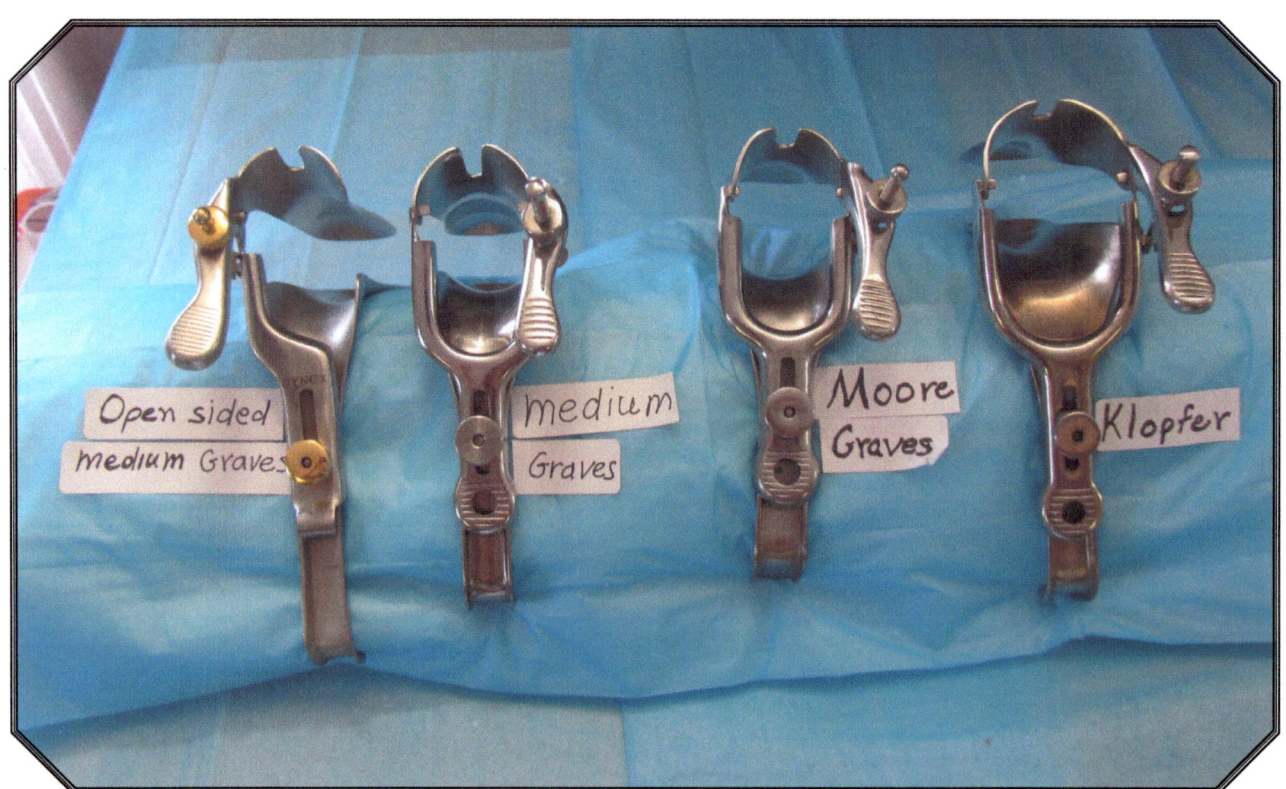

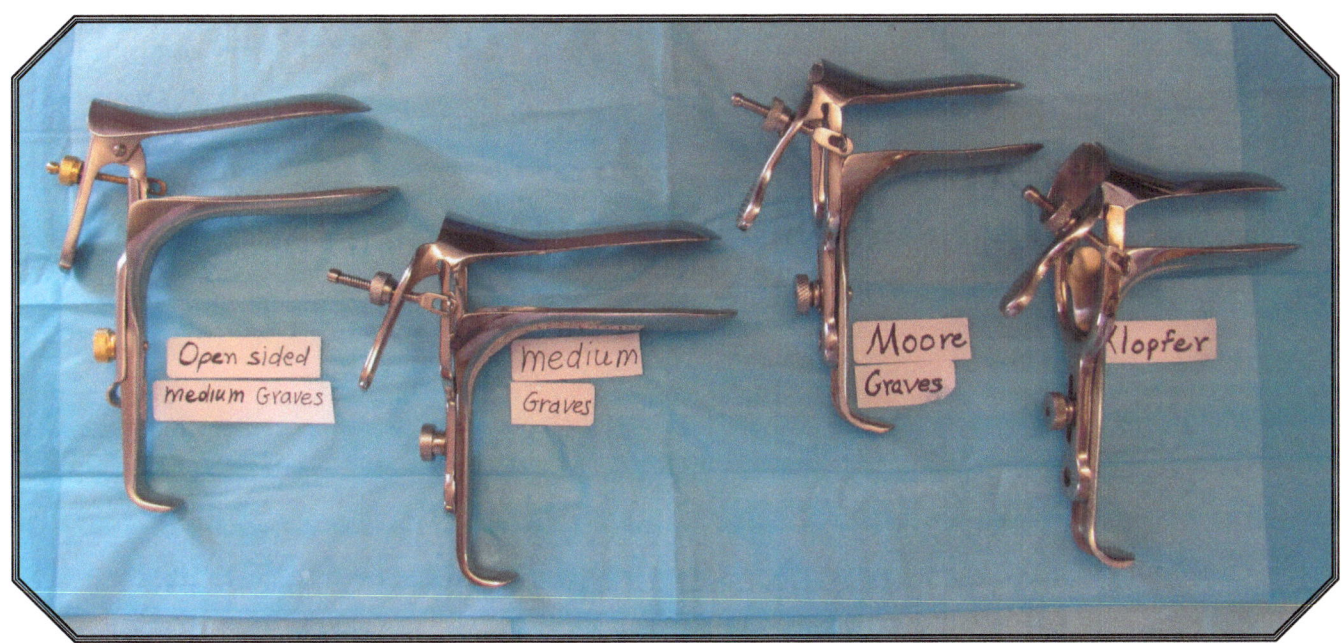

Routinely, I suggest utilizing a Moore-Graves speculum, which is shorter than the medium Graves. The average mobile cervix will descend when traction is applied, accommodating the shorter speculum. A longer speculum will limit your range in angling instruments anterior and posterior and increase the distance to the target part. For obese patients with a deep vagina, or those with decreased cervical mobility, the medium Graves is appropriate. The open-sided speculum can be useful in situations where one must remove and re-apply instruments. The Klopfer speculum has a larger opening than the others, allowing for more side-to-side manipulation. This is the preferred speculum for suturing cervical lacerations. Large Graves and large or medium Pederson specula are available and may be utilized to visualize the cervix as the patient's anatomy may require.

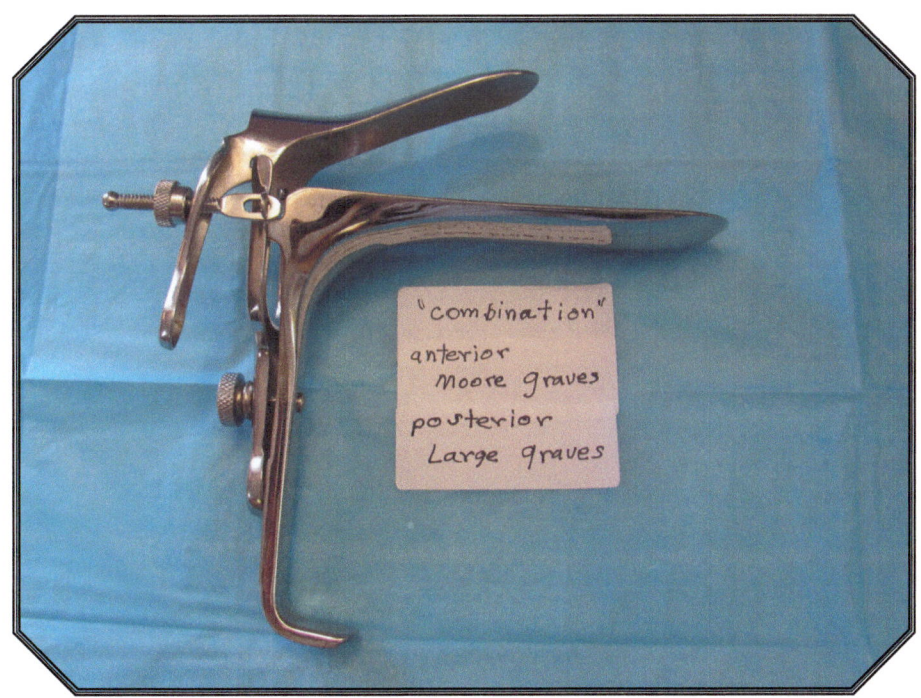

The bottom photo on the previous page shows a combination of Moore-Graves and large Graves, assembled by unscrewing the bottom bolt of both, then bolting the anterior blade of the Moore-Graves to the posterior blade of the large Graves. This is helpful when the cervix is fixed and retracted anteriorly. Another alternative is to remove the posterior blades from two specula and utilize both as one would Sims retractors. This requires an assistant to hold retraction.

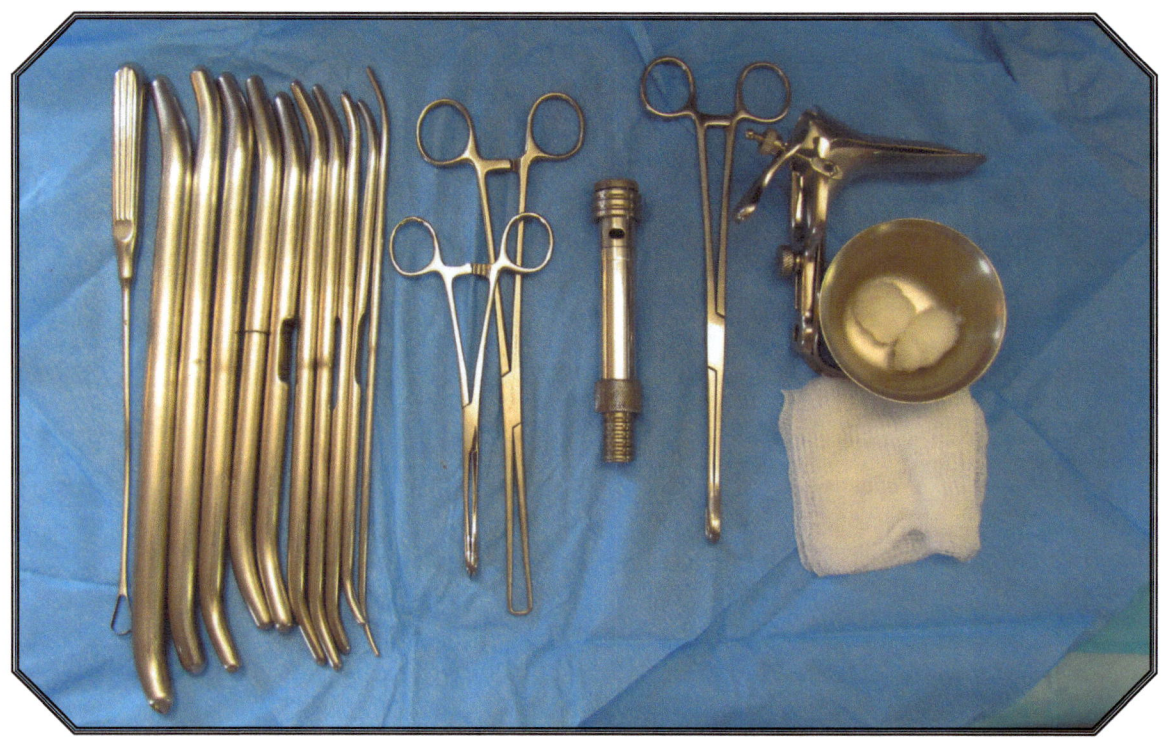

The standard D&E tray is above. Included is a curette, Pratt dilators up to #51, a small ring forceps, straight single-toothed tenaculum, handle for the 16mm suction cannula, large ring forceps, Moore-Graves speculum, bowl with prep solution, cotton balls and gauze sponges.

D&E forceps are described and illustrated on pages 10-12.

Operative preparation for extraction

Ask the patient if any dilators have fallen out. Be certain the patient is positioned properly before administering systemic sedation or anesthesia. If you have previously performed a pelvic examination, you may omit this step.

After selecting and inserting an appropriate speculum, laminaria should be removed with the ring forceps. In general, laminaria are extruding from the external os and can be removed quite easily. If laminaria are impacted, pull on the string or tip of a central one first to remove it, then remove one at a time. If it is extremely difficult to remove one laminaria, it may have partially perforated the cervix. Leave it in place and proceed with the evacuation; it will generally fall out as the cervix is dilated. If laminaria were not placed within the internal os, or fell out, it may be necessary to dilate the internal os or entire cervix with dilators.

After prepping, the anterior cervix should be grasped with an appropriate instrument. I generally initially utilize a single-toothed tenaculum. If the cervix is quite low, the smaller, shorter ring forceps is advantageous. As the cervix descends during the extraction, changing from a single-toothed tenaculum to a small ring forceps is sometimes helpful. Moderate traction should be applied while introducing instruments. Although not pictured in the models, the tenaculum or ring forceps is left in place for all maneuvers.

I prefer to administer vasopressin 4 IU in 4 cc of 1% lidocaine to reduce bleeding. If vasopressin is unavailable due to exorbitant cost, administer lidocaine plus epinephrine. Utilize 1% lidocaine with epinephrine 1:100,000, 4 mL plus 1% lidocaine, 16 mL, for a total of 20 mL. Ideally, this should be injected into the stroma of the upper cervix, at the level of the internal os. However, difficulties with visualizing the exact tip of the needle may result in inadvertently injecting into the amniotic fluid or cervical canal. Injecting submucosal in the lower cervix avoids this.

Ultrasound placement and suction

Next step is to place the ultrasound. Keep in mind that most assistants are minimally trained in ultrasound and you will need to direct them. One can argue transverse versus sagittal plane, but both directions should be utilized at times. Don't forget that ultrasound is two dimensional. If you are viewing in sagittal, it is possible to perforate to either side and be unaware. In transverse, perforations can occur above or below your plane, out of view. You must be able to view the forceps blades or suction cannula in relation to the part you are attempting to extract. This is generally more difficult with sagittal than transverse. Since I generally begin the evacuation in the lower uterine segment, transverse over the presenting part in the lower segment is my starting point.

A #14 cannula may be utilized for 13-14 wks. Fifteen wks. and above will require a #16. Steady the cervix with the tenaculum or ring forceps, apply traction and introduce the suction cannula into the cavity. In transverse, this appears as a small circle or line with a "shadow" deep to it. In sagittal, it appears as two parallel lines. Stop as soon as you see the suction tip in the cavity. Once you have the suction cannula in place, confirm that the ultrasound probe orientation is correct by aiming the cannula to one side of the uterine cavity and confirming that it does not appear on the opposite side of the scan. It is generally easy to change planes by directing the assistant to "move back" or "move toward me". I have utilized a model in lieu of ultrasound pictures to describe techniques because it is often quite difficult to explain and visualize from still ultrasound views. Instruments can be easily identified in real time by moving them slightly.

In earlier gestations, placing the suction opening in direct proximity to the calvarium, and then closing the O-ring may result in partial or total removal of fetal parts and eliminate the need for use of forceps. For more advanced gestations, the evacuation begins by suctioning off amniotic fluid. Close the O-ring after you have confirmed that you are in the cavity; then immediately open it as soon as you see fluid. This hopefully keeps the cannula from being occluded by the cord or presenting part. If fluid is escaping around the cannula, withdraw to the point where you can suction off most of the fluid, then insert again. As noted above, check ultrasound orientation. Stop when most of the fluid is removed on ultrasound.

Federal law prohibits intentionally extracting either the fetal trunk past the umbilicus or the fetal calvarium to outside the patient's vagina and then inducing fetal demise. (2) To avoid any question of non-compliance, bring the umbilical cord through the cervix and sever it as early in the case as possible. This is generally easily accomplished while removing fluid at the beginning of the case, utilizing the suction cannula.

Removing the speculum while leaving the cervical tenaculum in place

At times it is necessary to remove the speculum in order to open the forceps widely enough or remove a large part. (See next page for explanation of the photo below.)

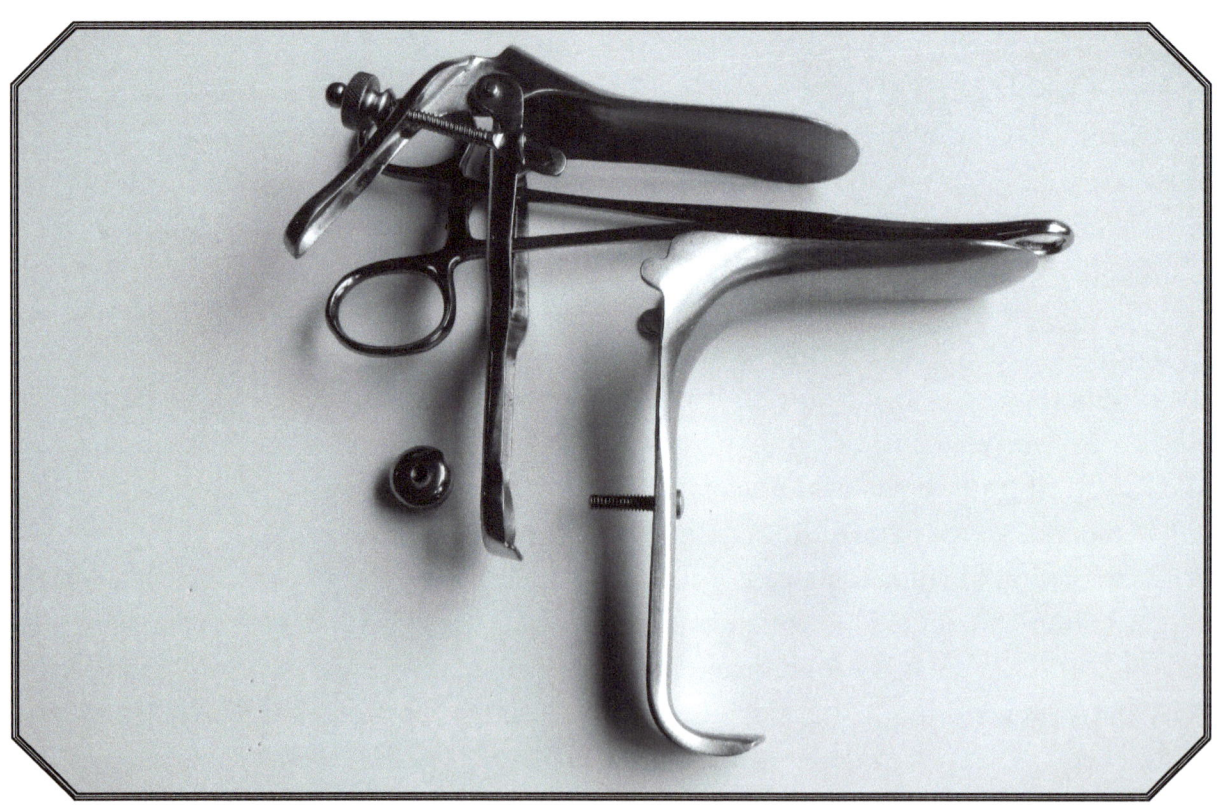

To remove the speculum while leaving the tenaculum or ring forceps in place, open the top screw fully. Completely unscrew the bottom bolt, separating the anterior and posterior blades. Remove the posterior blade first.

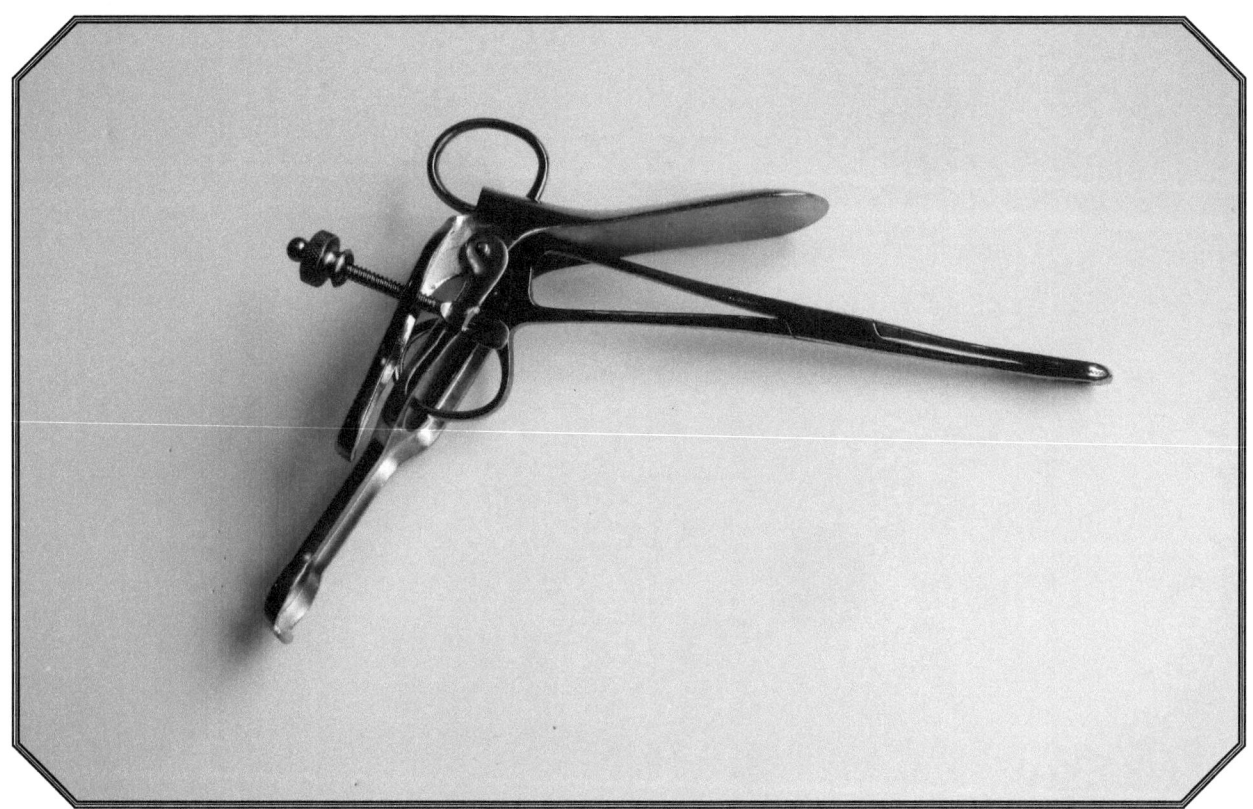

Remove the anterior blade by utilizing the notch (as seen on page 4) to gain enough space to slide the speculum opening over the handle of the tenaculum or sponge forceps while leaving it on the anterior lip of the cervix. The posterior blade can then be replaced as needed to visualize the cervix.

The next step is to select the forceps.

Abortion forceps

Abortion forceps are an interesting meld of Obstetrical Forceps and grasping forceps from other disciplines. They all consist of a handle, shank, hinge or lock and blades (or jaws). Some are straight; others have various degrees of a "pelvic curve". They are available in a variety of lengths. Many look quite similar. Since I "inherited" these, I am unsure of their exact names, but for the purposes of this book, they will be referred to as labeled in the photos below. I utilize the small Sopher, medium Sopher, 11-in. Bierer and 13-in. Bierer.

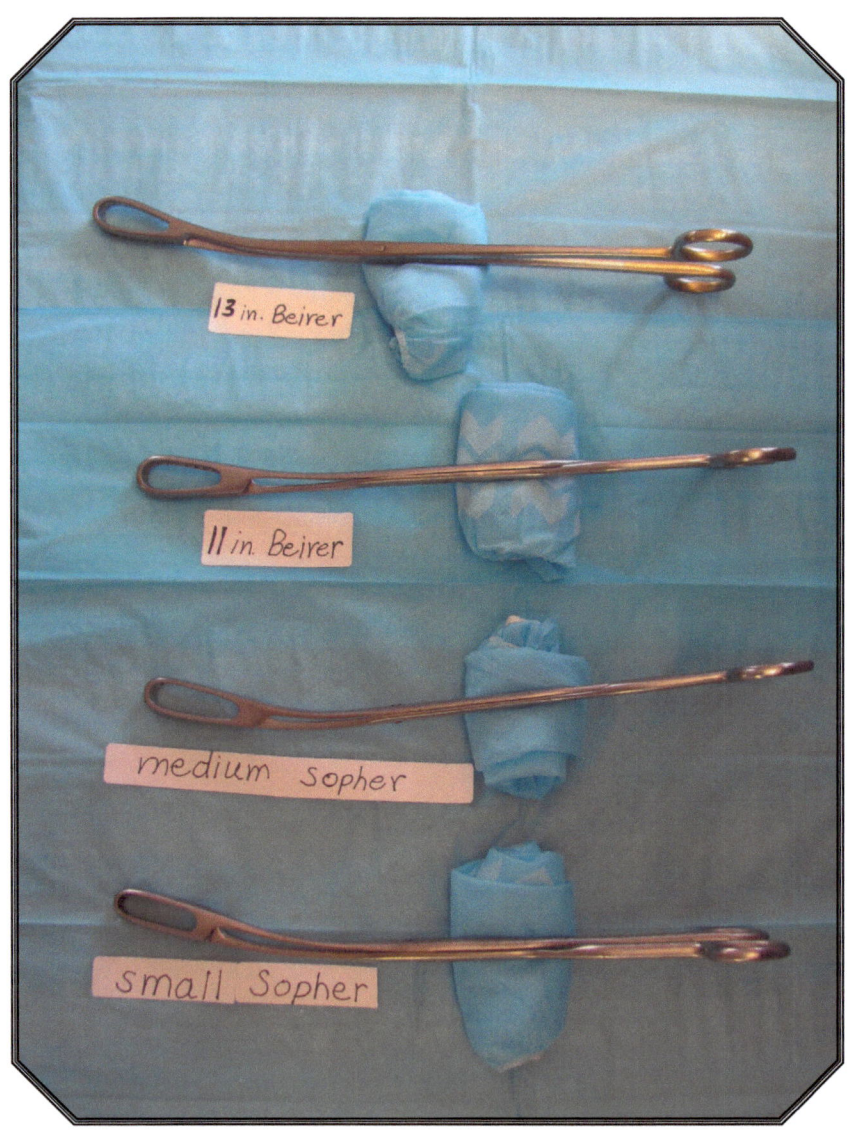

It is quite important to take note of the "pelvic curve". One encounters many situations where it is necessary to negotiate a curve in the uterus or pelvis, for instance, if one is extracting a fetal part from the cornua, or from the fundus in a very anteverted or retroverted uterus. For these situations, it is necessary to utilize a curved instrument. Only the 11-in. Bierer is a relatively straight forceps.

In the side view on the previous page and the view from above next, note the difference in the blades (jaws). The Sopher forceps have smaller blades with an oval shaped fenestration. The Bierers have a teardrop-shaped fenestration. The small Sopher can therefore be introduced in a situation where cervical dilation is less, and fetal parts are smaller. It is a good forceps for a 15-16 week case, and will pass through a cervix dilated to a #43 Pratt dilator. The Medium requires more dilation but is still useful in 16-17 weeks. Occasionally one encounters situations where the gestation is more advanced, but the cervical dilation is inadequate, for instance, if the laminaria have fallen out or were not inserted far enough. These cases can often be managed by extracting the small parts with a Sopher initially, changing to a larger forceps when the cervix dilates farther.

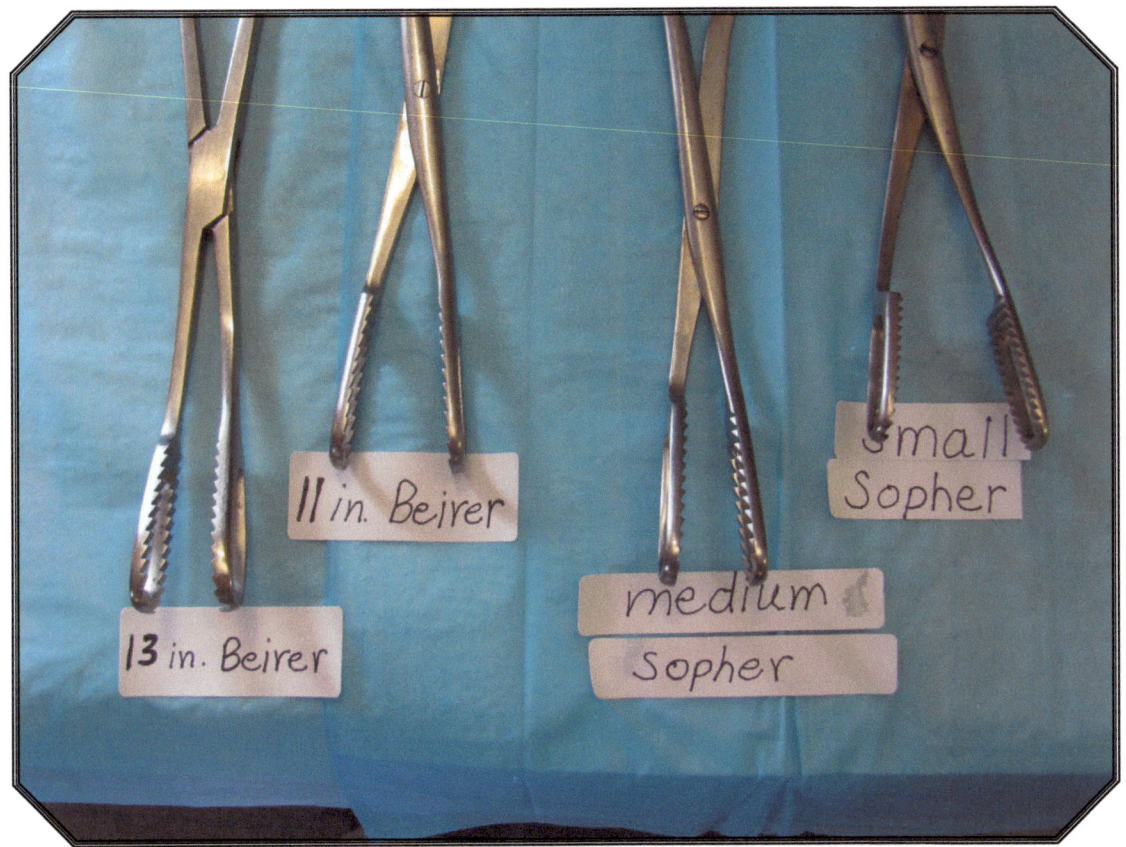

Note that the blades on the Bierer are heavier and the teeth are larger. Hence, the Bierer is a better grasping forceps. Also note above and on the next photo that the 13-in. Bierer has a straight, heavier shank and more secure lock. This is the only forceps pictured which can be easily utilized to compress large fetal parts.

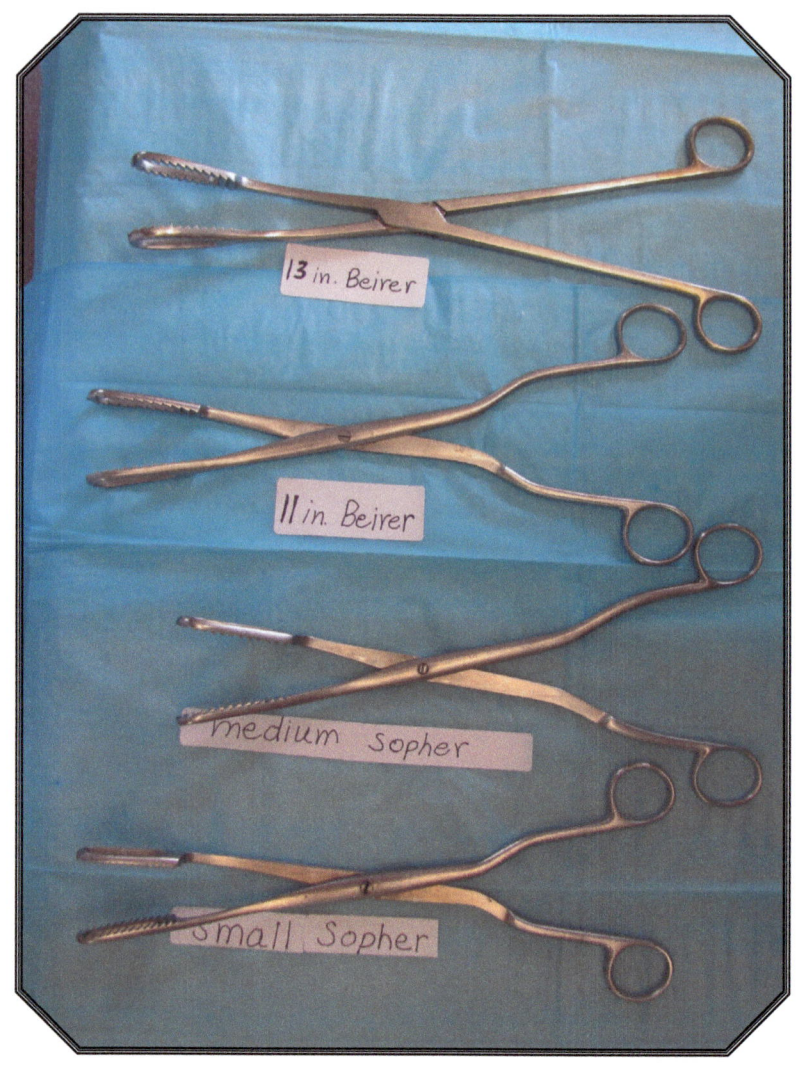

When selecting the initial forceps, consider both length and curvature. Attempt to visualize the distance between the presenting part(s) and the opening of the speculum. The longer length instrument can work against you by decreasing your ability to open the instrument. If the cervix is "high", or the presenting part is distant, for instance with a transverse lie, the longer instrument will be required. Curved forceps have an advantage in negotiating curves. The disadvantage is that, when you rotate the forceps, the blades swing through an arc as opposed to simply rotating. I generally start with the 11-in. Bierer, if the presenting part is relatively accessible and low in the pelvis.

In general, forceps (and most other instruments) should be held over-handed. The range of motion of your wrist is greater in this direction. The exception is a situation where you are attempting to grasp a part anterior to your forceps or from the patient's right or need extreme flexion of your wrist. If you are utilizing a curved instrument, utilize the curve to your advantage.

Instructions for all maneuvers that follow are for right-handed individuals. Left-handed operators will need to transpose them accordingly.

Models for demonstration of maneuvers

Below are photographs of the models I am utilizing. Labels, the cervical tenaculum or ring forceps and the speculum have been omitted from most views to facilitate demonstration of the instrument, fetal and hand positions.

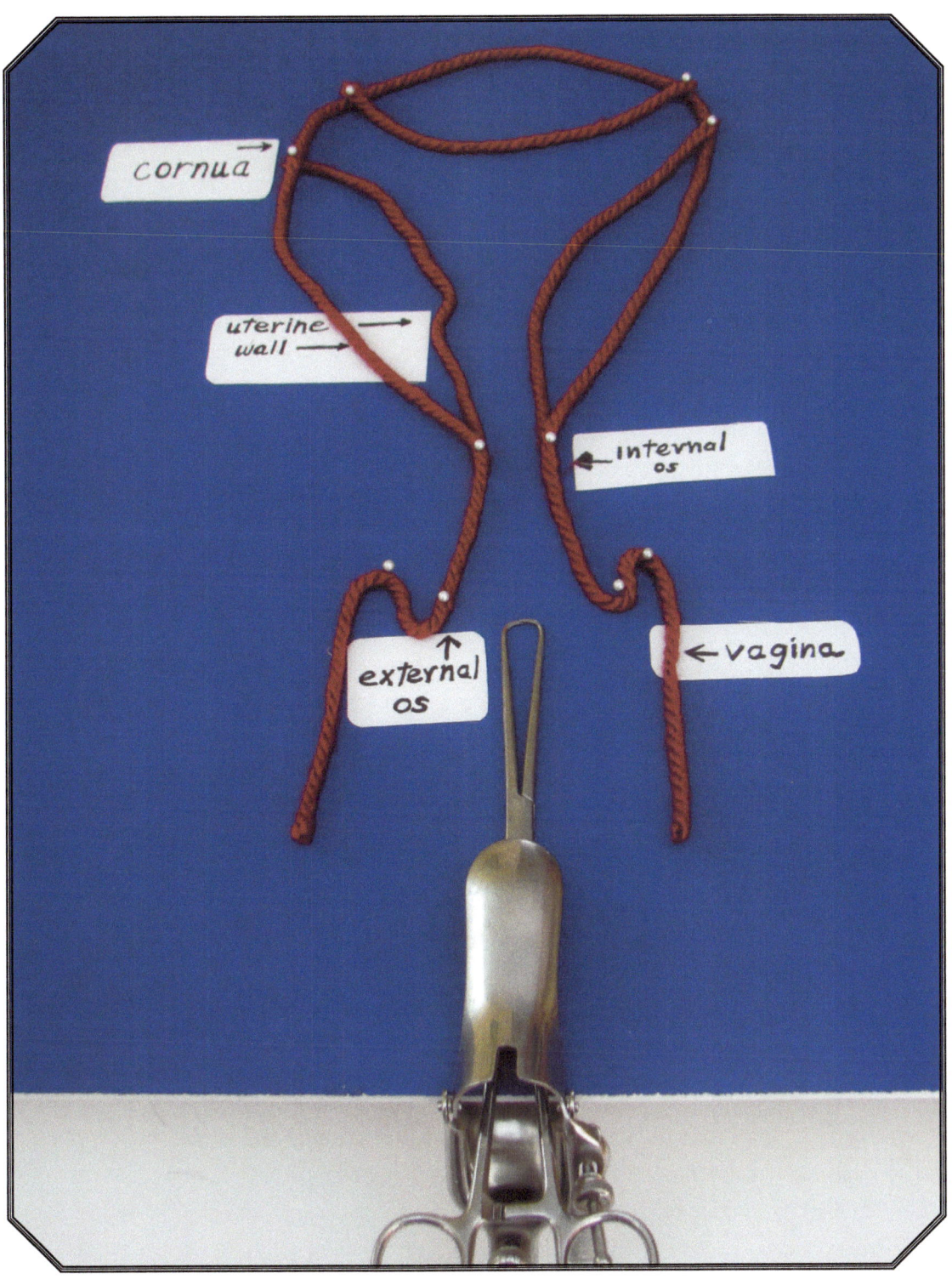

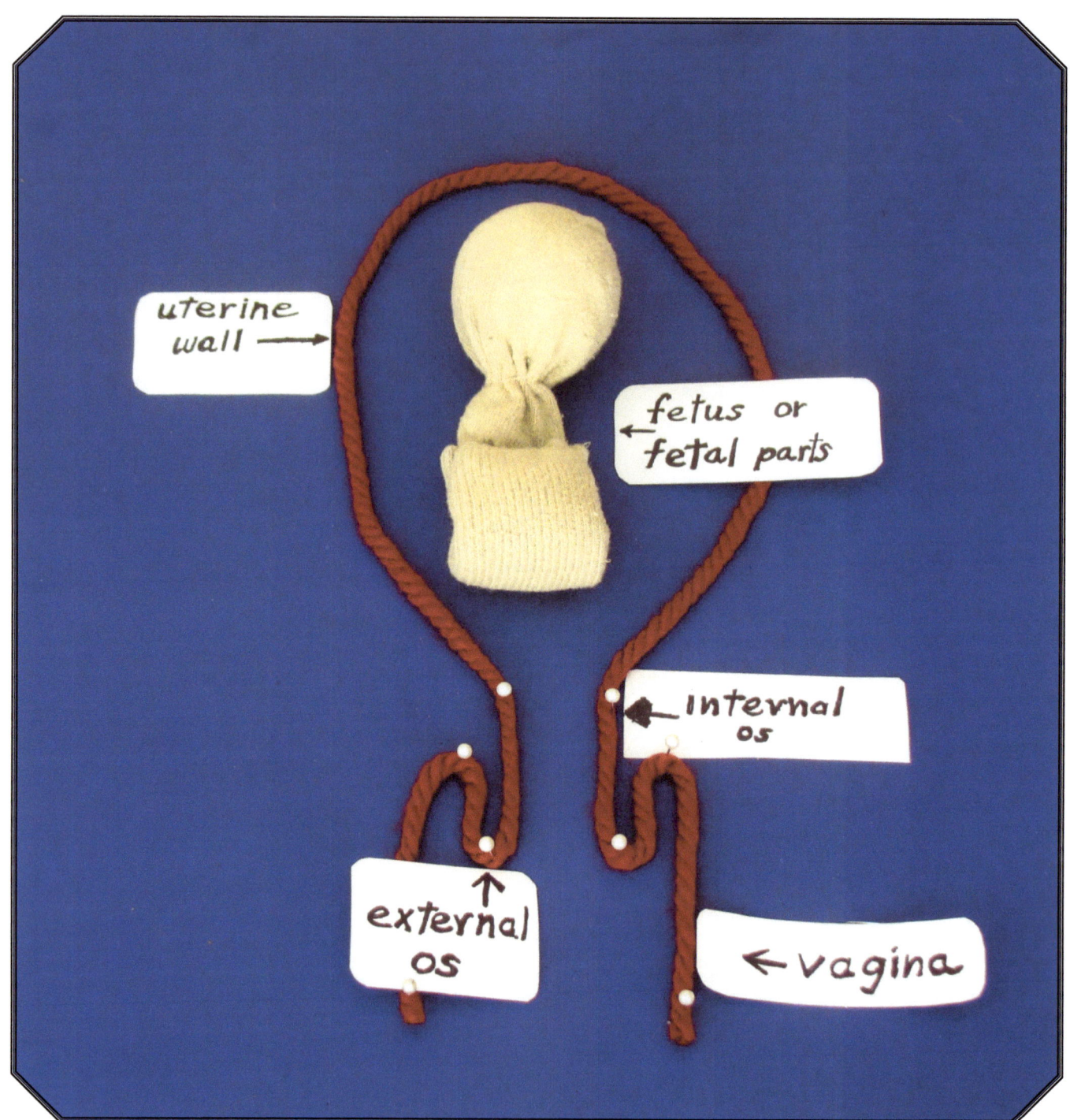

Operative plan

In the maneuvers described subsequently, visualize the forceps on ultrasound when it is first introduced into the lower uterus, shift planes to visualize the part you are attempting to grasp, then advance the forceps while closed until you can visualize them. Do not open the forceps until you visualize them in apposition to the target part. Utilizing sagittal ultrasound is acceptable, but you must see both the part you are attempting to grasp and your forceps blades. Forceps blades appear as two parallel bright lines or four dots when they are opened slightly. Open and close slightly, or move right or left if you are not sure.

It is sometimes helpful or necessary to have an assistant apply fundal pressure. The assistant holding the ultrasound is generally asked to do this. The operator's left hand can also be utilized to palpate and steady the fundus, or bring fetal parts into contact with the forceps (Hanson maneuver). If you encounter difficulty attempting to remove the calvarium first in a cephalic presentation, it may be more feasible to start with the shoulder or upper extremity. Breech presentation extraction is generally initiated with extraction of the lower extremities. A standard breech extraction, as described in obstetrical textbooks, can be utilized as soon as the cervix is sufficiently dilated. In a transverse lie, it is sometimes possible to convert to a breech presentation by identifying the leg and pulling it down with the forceps. Converting to a shoulder presentation by pulling down the upper extremity is another option. These maneuvers will be described and illustrated subsequently.

Description and illustration of basic extraction techniques

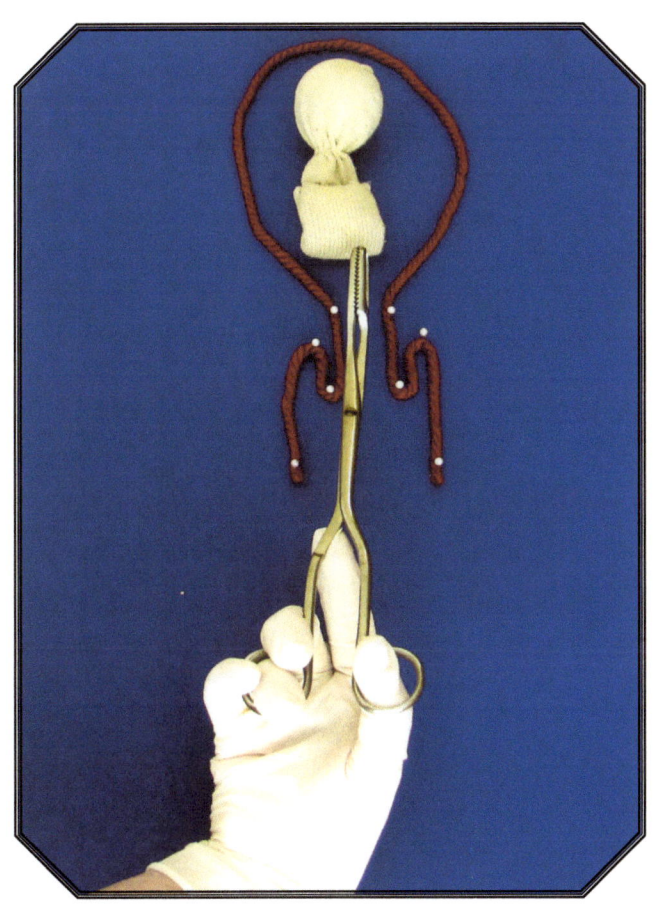

Apply traction on the cervical tenaculum or ring forceps to stabilize and straighten the cervix.

To begin the extraction, introduce the forceps, closed, into the cavity, directly anterior or posterior, utilizing the curve to the direction of the cervical canal. If the uterus is anteverted or straight, place your hand underneath the forceps with the curve pointing anterior.

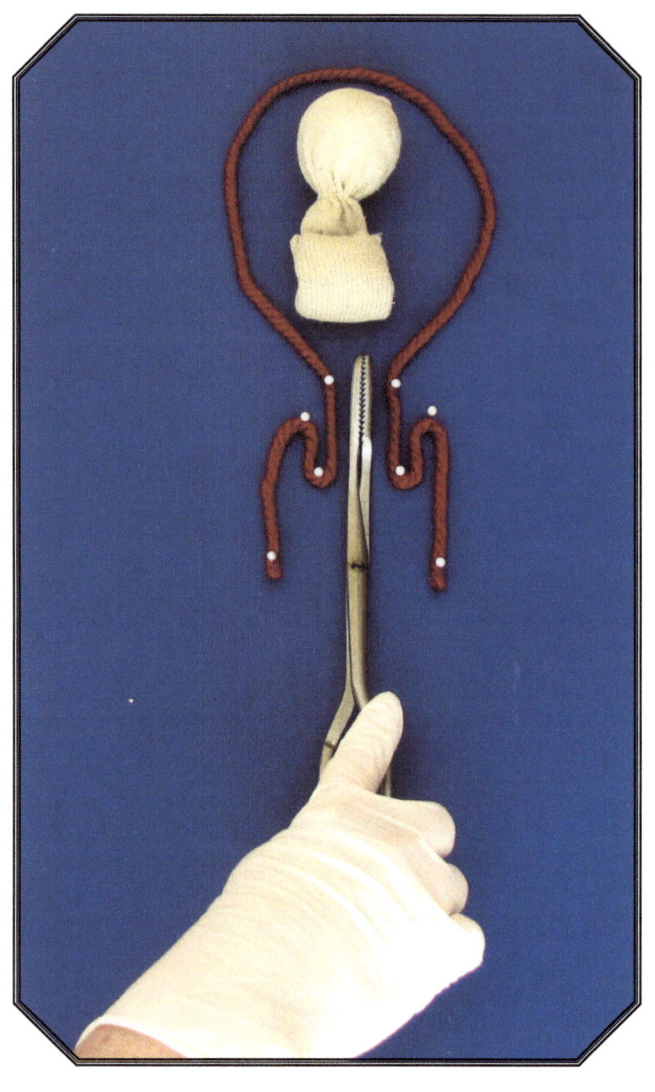

If the uterus is retroverted, turn your hand 180 degrees to aim posterior.

Both anterior and left lateral approaches that follow allow you to rotate or otherwise "aim" the forceps utilizing multiple variations in direction to grasp and manipulate the part. Approaches from the patient's right (page 26, last photo page 35), posterior or from below (page 22) can also be utilized, but because your wrist and arm have less range of motion from these positions, may be more awkward and less effective.

Approach from lateral

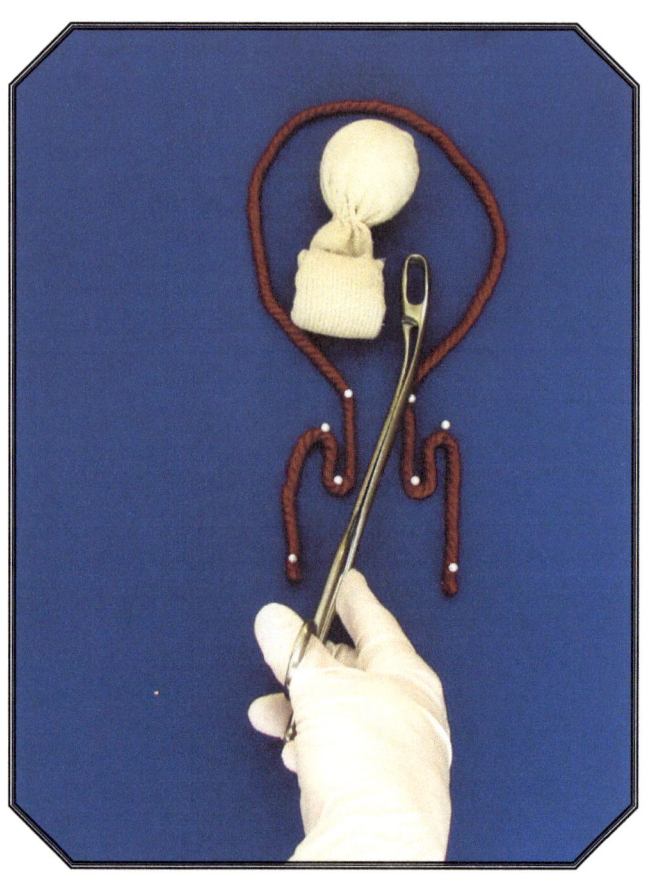

After clearing the internal os, confirm intrauterine location on ultrasound. Position the forceps blades to the patient's left of the part you are targeting. Turn the forceps so that the curve points to the patient's right, with your hand on the outside of the forceps.

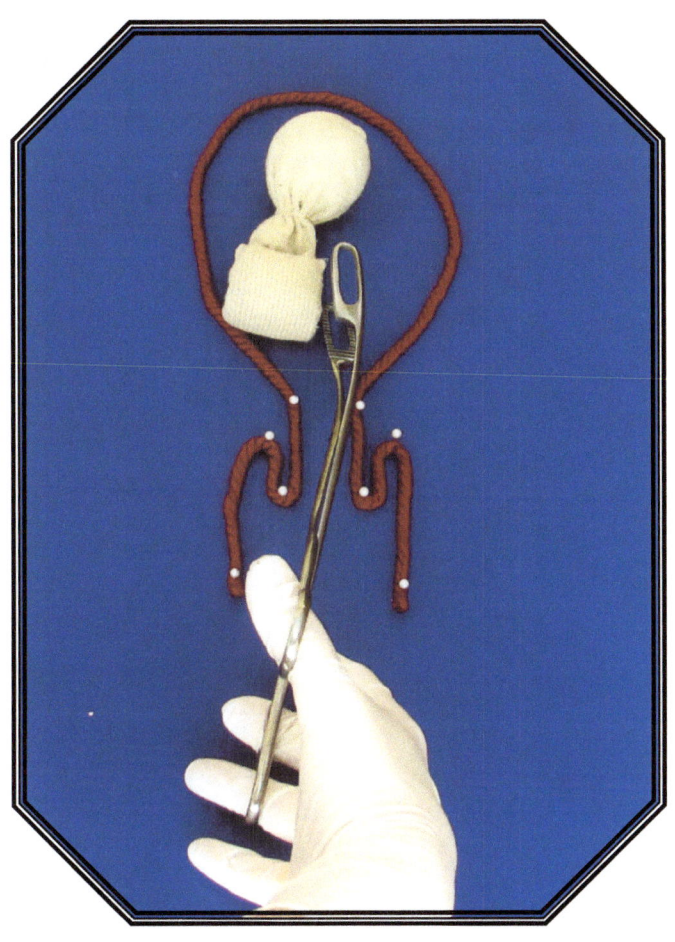

Open the forceps widely enough to encircle the part. This may require some steady, gentle force if the uterus is contracted or uterotonic drugs have been given. The alternate hand-forceps position (last photo page 22) may be utilized. You should see both blades of the forceps on the ultrasound screen, one anterior and to your right of the target part and one posterior and to your right. Move the forceps slightly to confirm your placement and visualization.

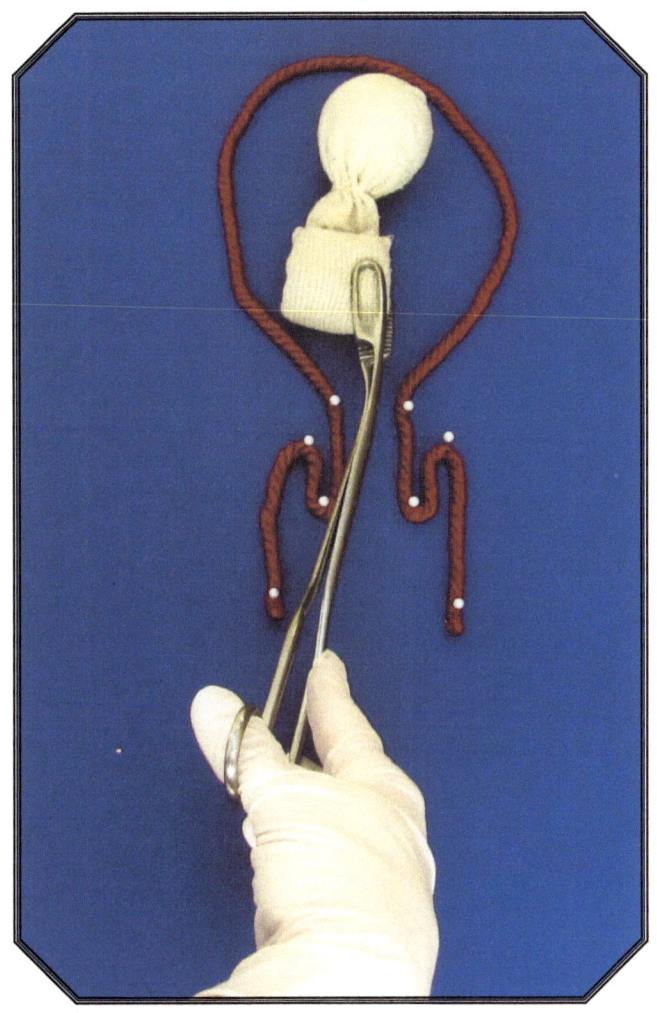

Advance over the part and grasp. If you "miss" and grasp a portion of the part or another part, proceed to extract what you have grasped. It is generally easy to remove the placenta if it has been grasped (page 36).

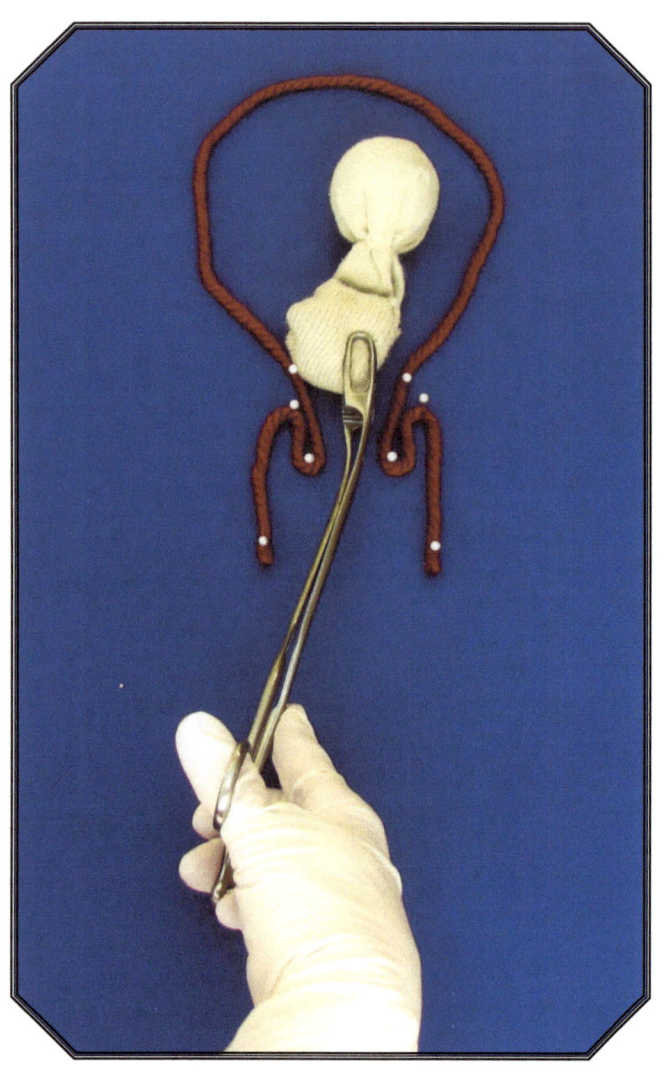

Bring the part down to the cervix for extraction.

If the part does not descend or otherwise move easily, determine that you have not inadvertently grasped something other than the fetus, such as a uterine septum. Maintain your grasp but stop pulling. Gently rotate the forceps clockwise, then counterclockwise, and confirm with ultrasound that the fetus or part moves with the forceps freely before proceeding.

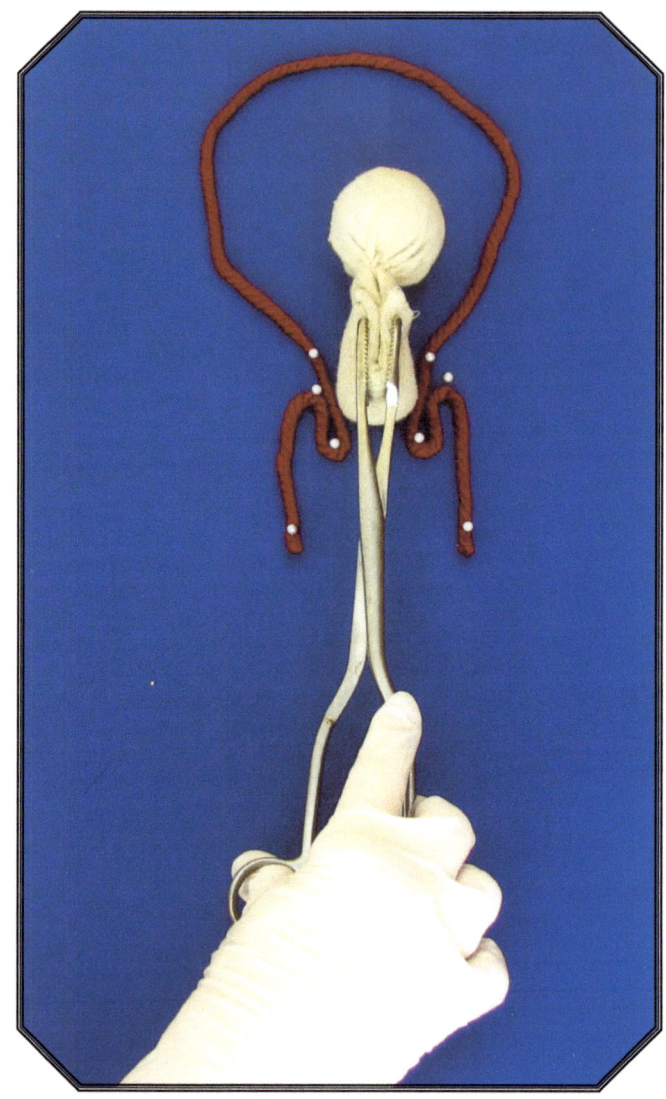

With earlier gestations and adequate cervical preparation, continuous gentle to moderate traction in a straight outward direction will result in the part descending through the cervix ...

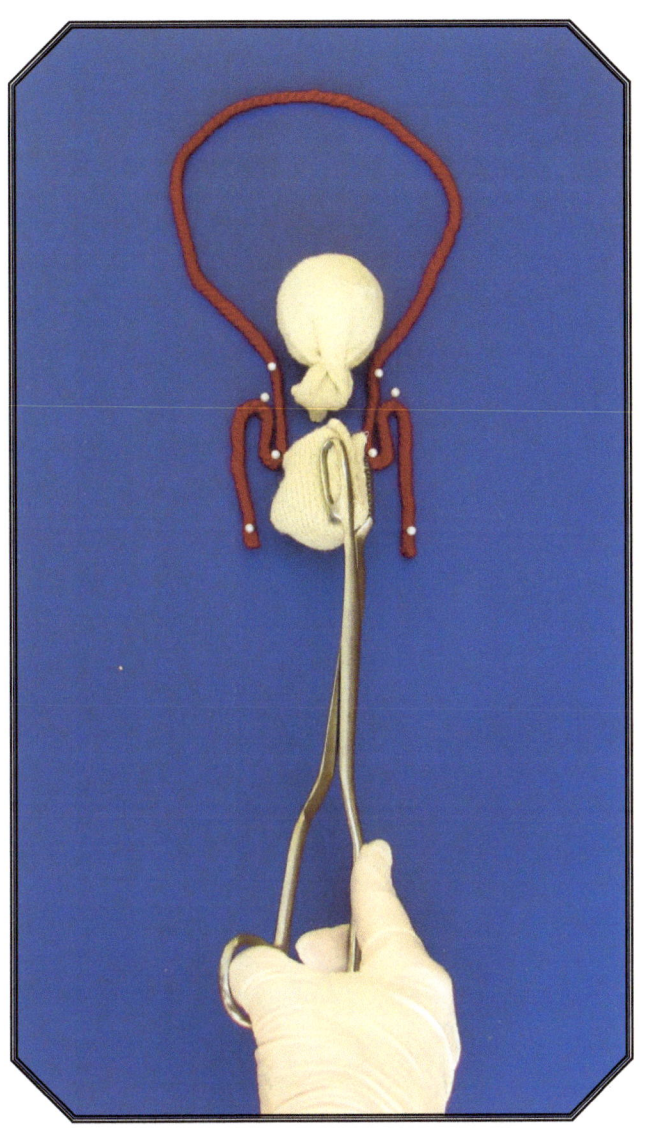 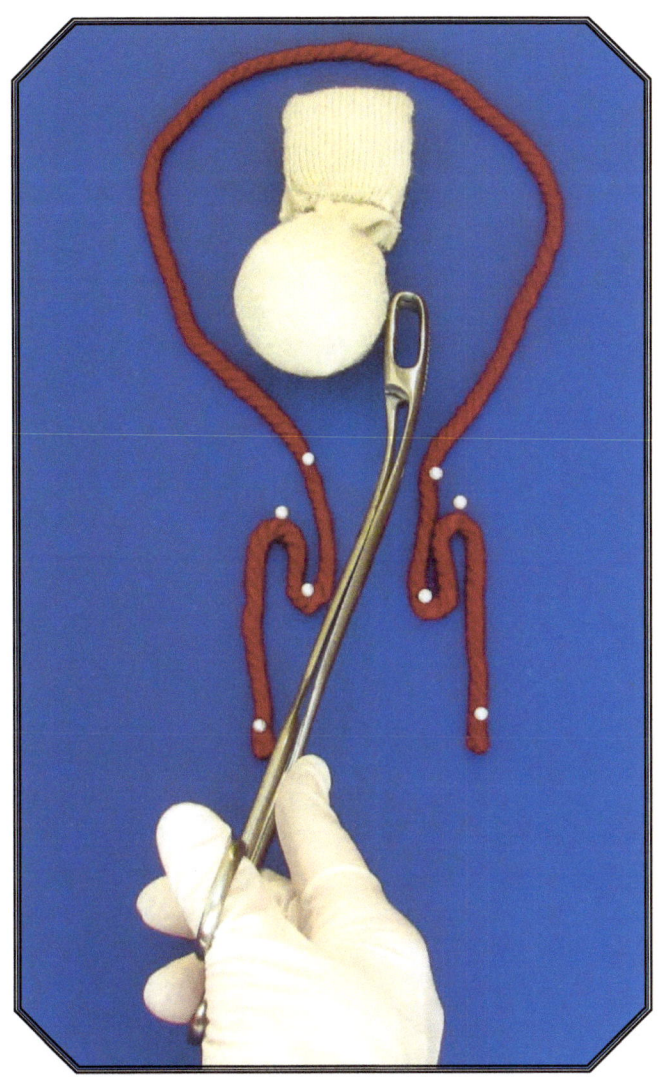

... and then separating. Excessive traction may result in lacerations of the internal os and cervical canal. Do not lean backward, brace your foot against the table or pull with both arms. If moderate traction is not effective, utilize the "turn and pull" technique described on pages 28-29.

A similar approach may be utilized to remove the calvarium. Locate the largest diameter on transverse ultrasound and move to a plane slightly above (toward the patient's head). Place the forceps blades to the patient's left of the calvarium, with blades pointing to the patient's right. Open the forceps widely enough to encircle the calvarium. Again, this may require some pressure. Confirm on ultrasound.

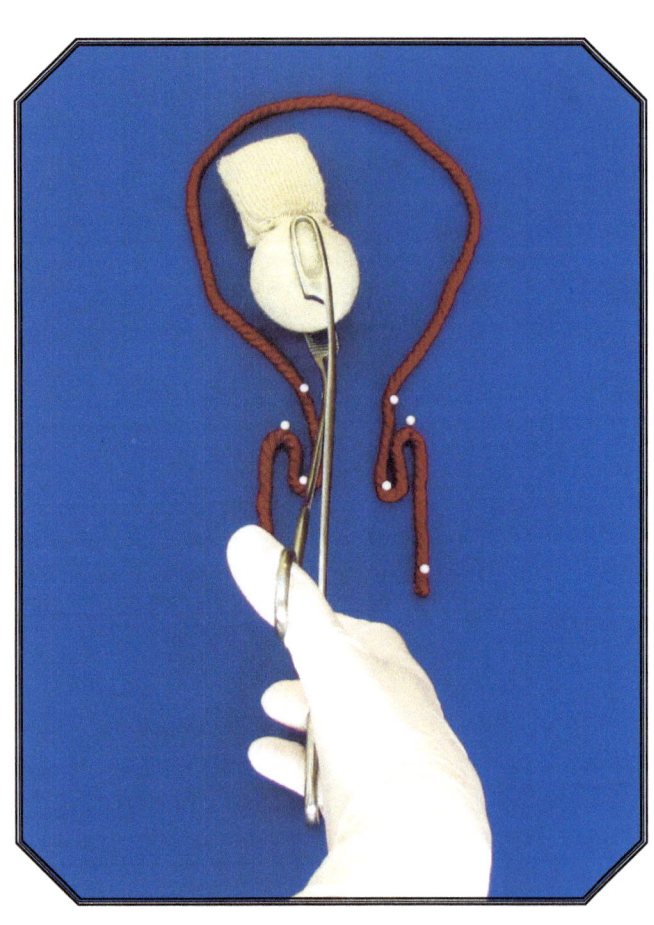

Advance the forceps over the calvarium. Close the forceps.

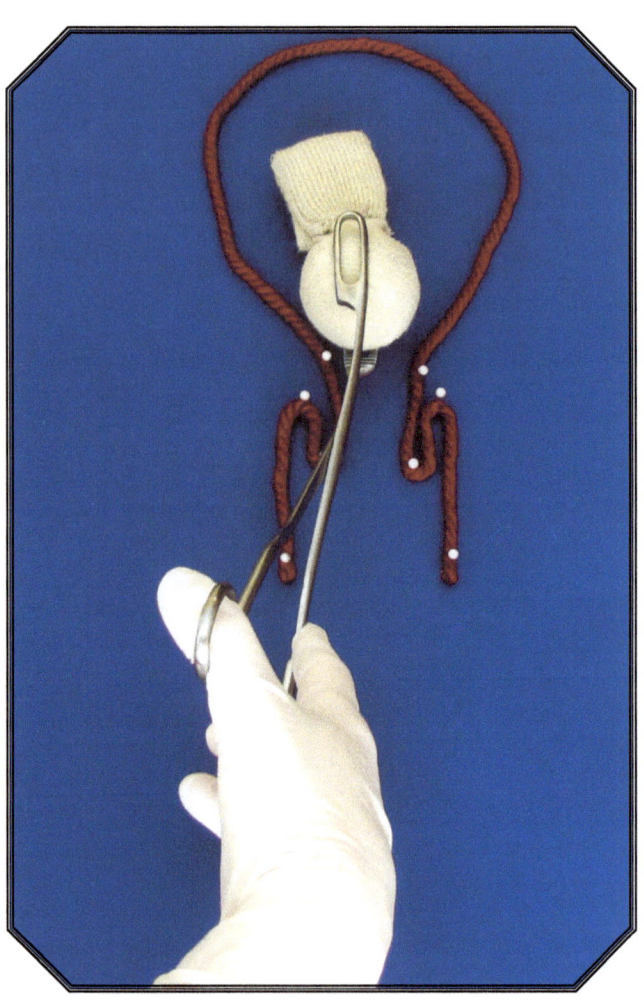

Bring the calvarium down to the internal os. (Next step on page 22)

Approach from anterior

Since the calvarium can be quite "slippery", an alternative is to approach the calvarium from anterior. Approach from anterior may be helpful in extracting other parts also.

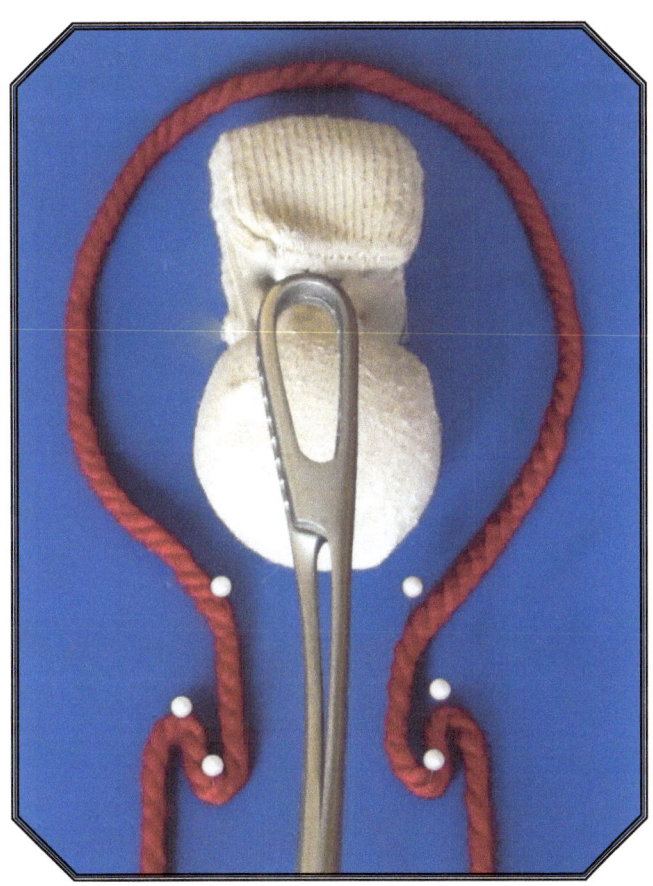

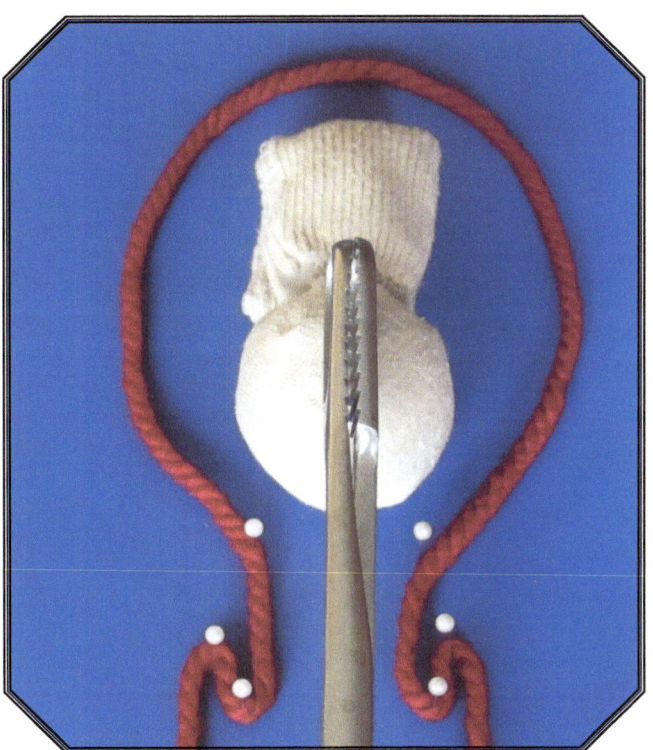

Turn the blades to point posterior.

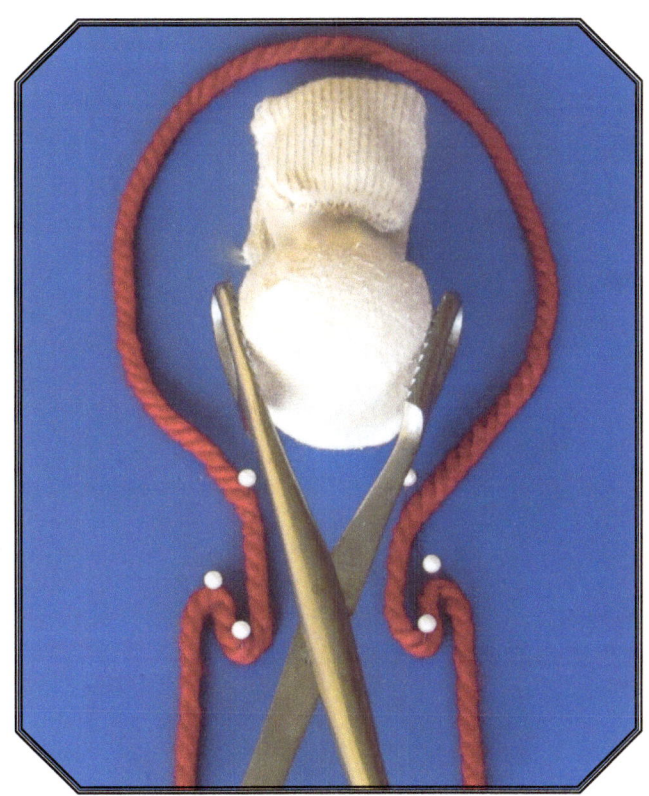

Again, introduce the forceps closed. Position them anterior to the calvarium and just above the largest diameter.

If the uterus is very contracted, and it is difficult to maneuver, approaching from the patient's left (page 19, second photo), then positioning to direct anterior may be effective. Fundal pressure to steady the uterus may be required.

Open, advance the forceps posterior, and grasp. Bring the part to the internal os.

(Next step for both approaches)

Once the calvarium has been grasped and brought to the internal os, begin compression. Exert some traction while you are compressing to compress against the cervical os.

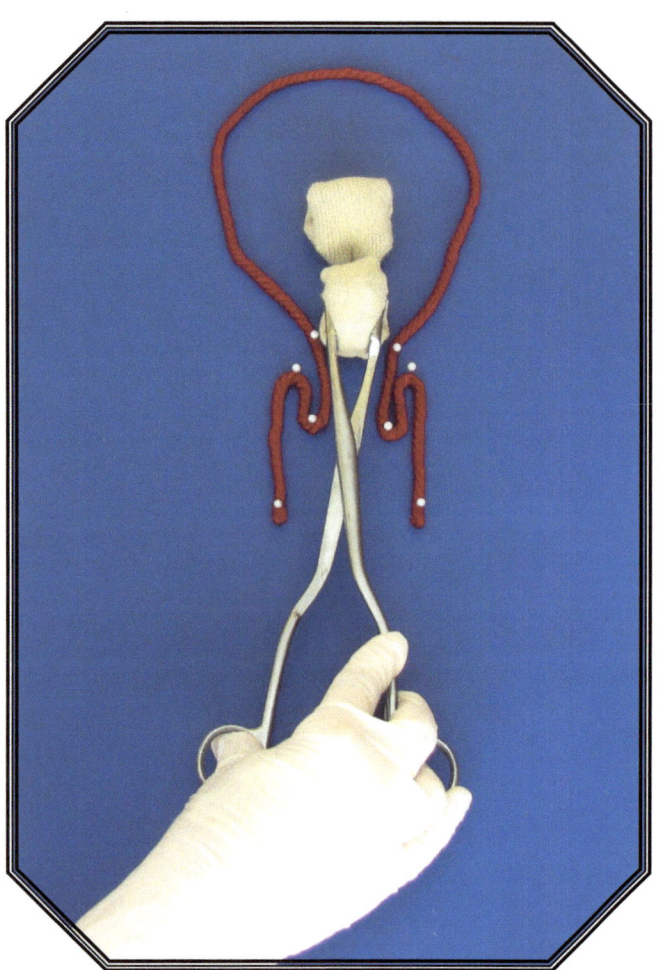

White matter may appear at the cervix, confirming decompression. If decompression is not accomplished easily, utilize the maneuver on page 33. The calvarium can then generally be easily brought through the cervix. If serious resistance is encountered, utilize the turn-and-pull technique described on pages 28-30. In general, any fetal part remaining follows easily, but if not, it can generally be grasped just above the internal os utilizing an approach from below (to follow) and extracted. If not, utilize a lateral or anterior approach.

Approach from below

This approach is best for situations where the target part is firmly applied to or wedged into the cervix.

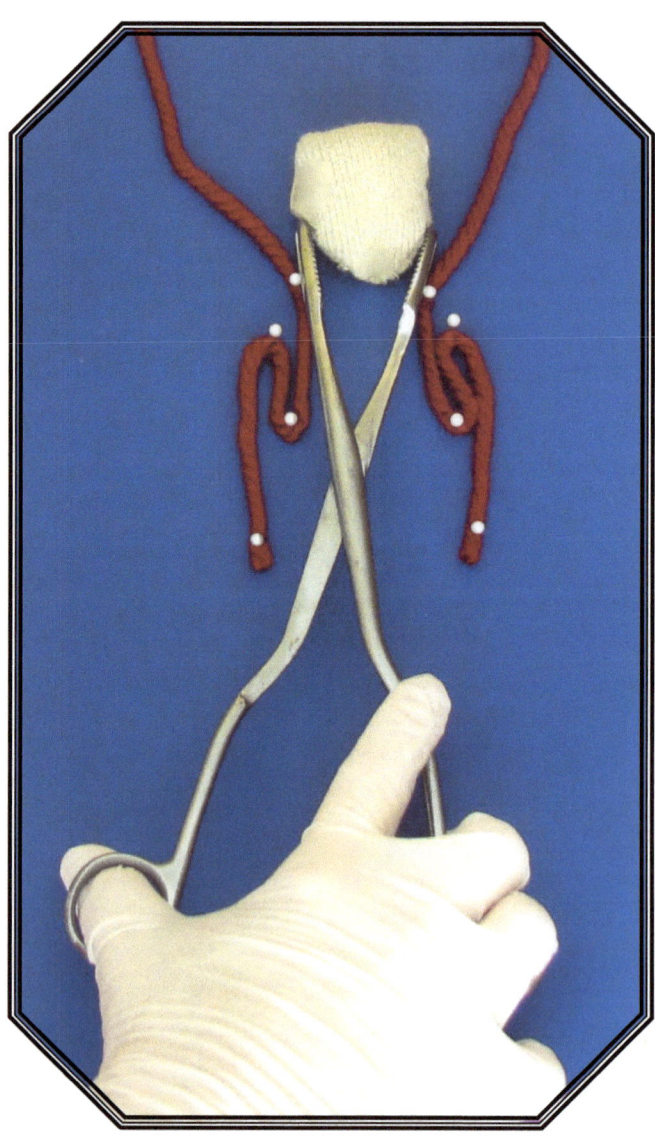

In this approach, begin to open your forceps as you traverse the internal os, and, visualizing the part and forceps blades with ultrasound, open widely enough as you advance to grasp the part. Grasp and extract as previously described.

Management of "slip"

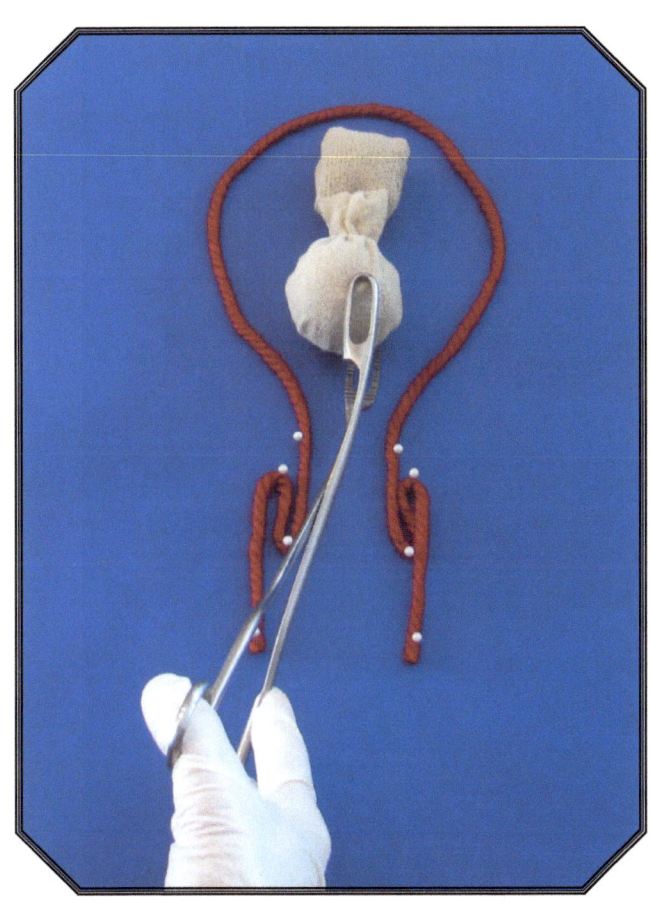

In the above, the operator may feel the calvarium has been securely grasped.

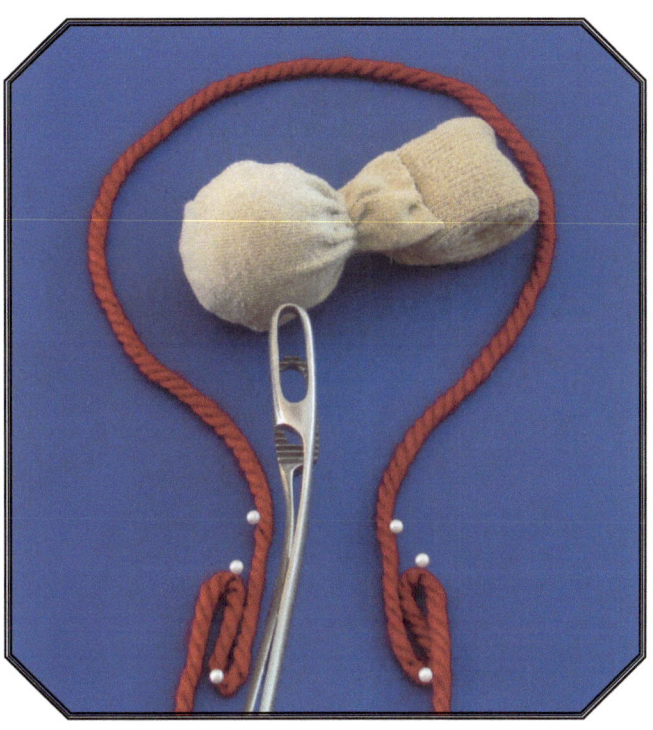

It "slips" when the forceps are closed. Placing the forceps slightly past the largest diameter of the calvarium, as shown on the next page with an anterior approach, aids in preventing the calvarium from "slipping away" and moving into the fundus or to the patient's right during grasping or compression.

Management of transverse lie

Conversion to breech

Transverse lie with the back down is the most challenging fetal presentation. I have demonstrated both conversion to breech and conversion to shoulder presentation with extraction. These extraction maneuvers may be utilized in other situations also.

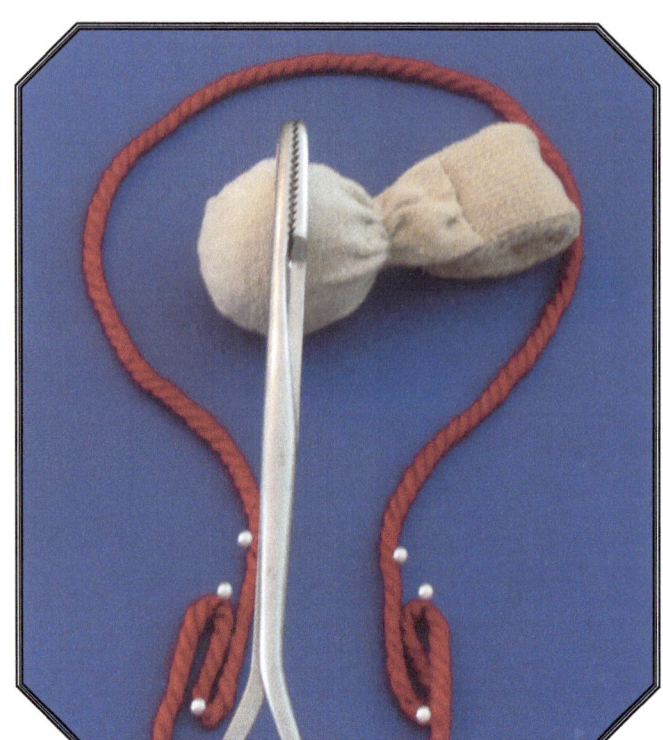

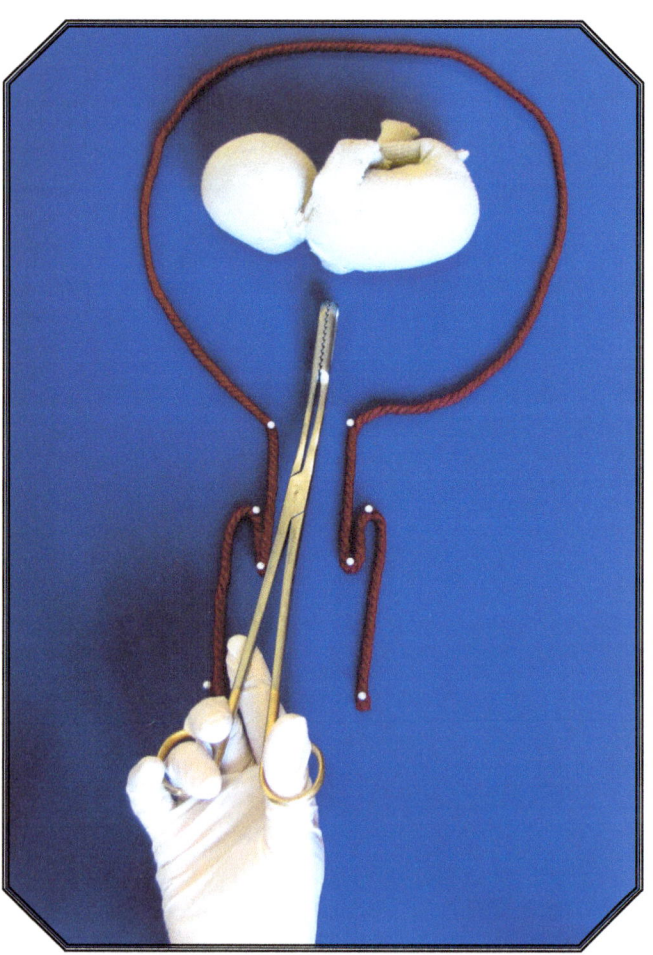

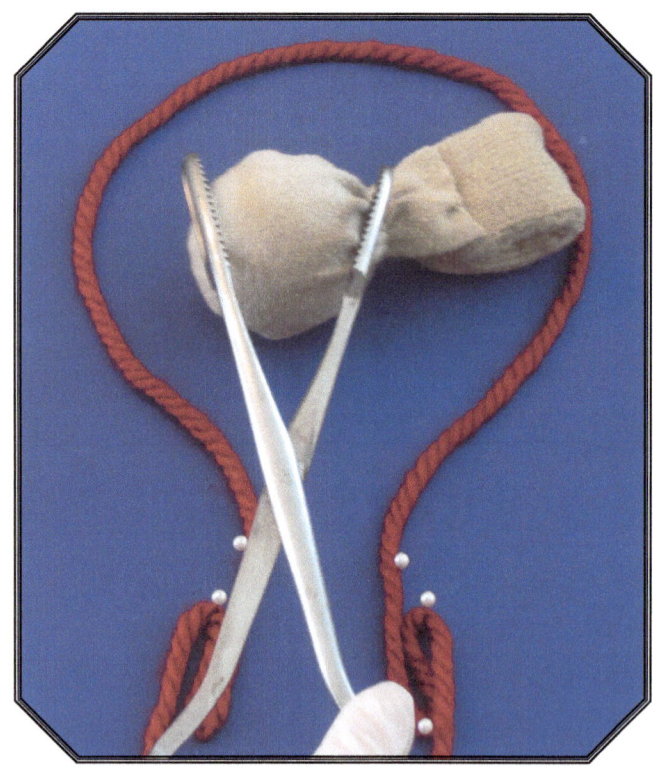

If a "slip" happens, a second attempt at grasping utilizing a different approach may be successful. If unsuccessful, it is generally best to target another part.

For grasping the lower extremity to the patient's left, introduce the forceps with the curve pointing anterior if the uterus is anteverted; posterior if retroverted (page 15-16). Confirm intrauterine location with ultrasound. Ask an assistant to apply fundal pressure or utilize the Hanson maneuver (not shown).

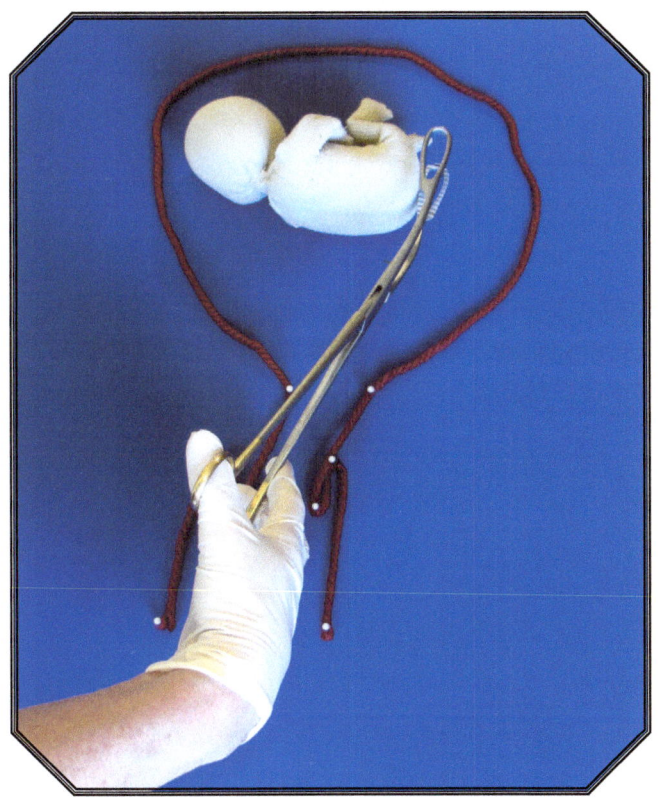

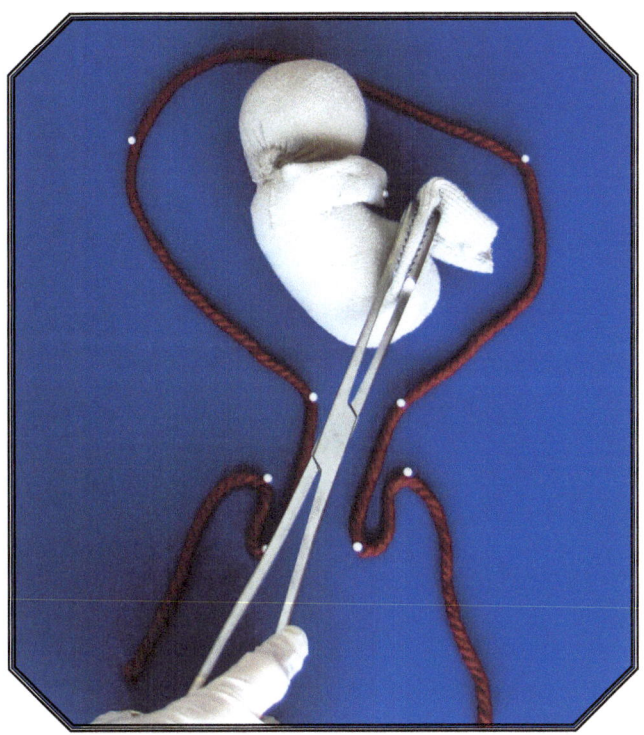

Grasp.

Position the forceps to the left of the fetus. Locate the extremity and your forceps on ultrasound. Open the forceps.

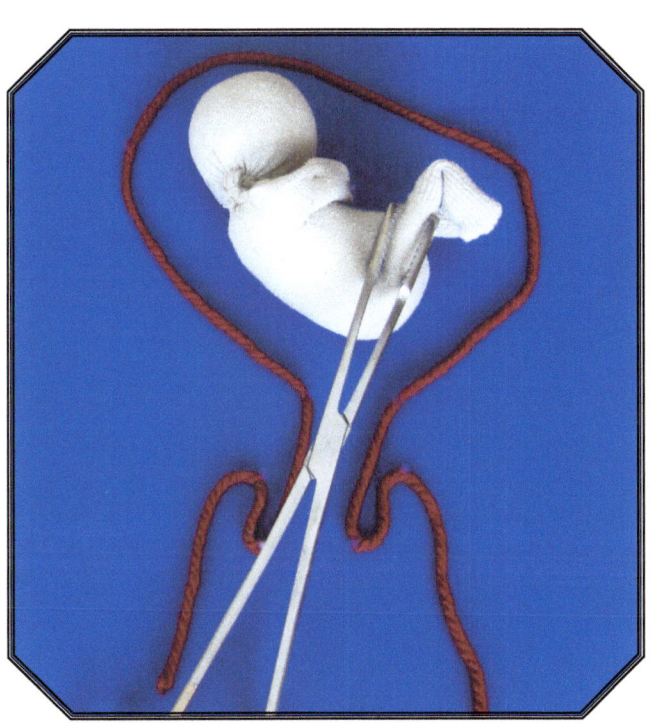

Advance the forceps, rotating anteriorly if necessary.

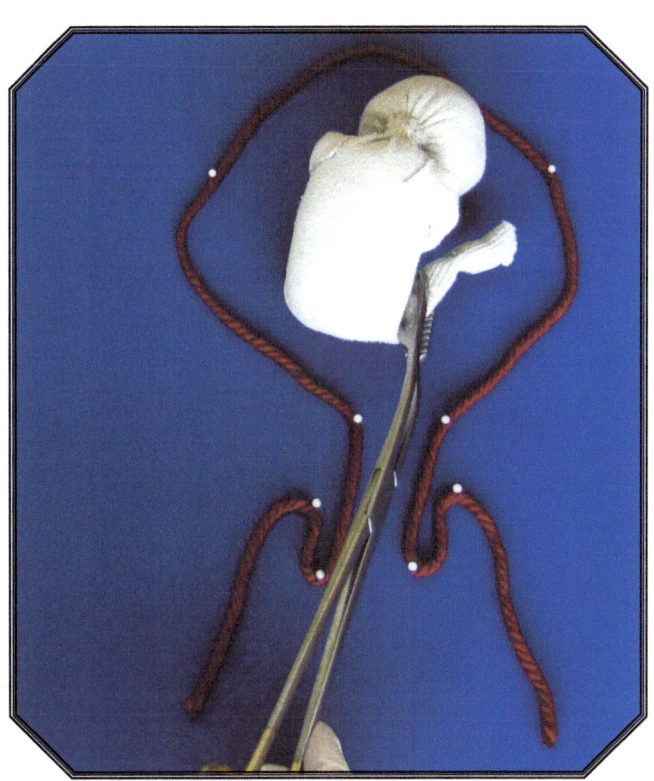

Pull the extremity down while turning clockwise. It may be quite feasible to extract as a frank breech. Before extracting as a breech, be certain that the umbilical cord has been severed.

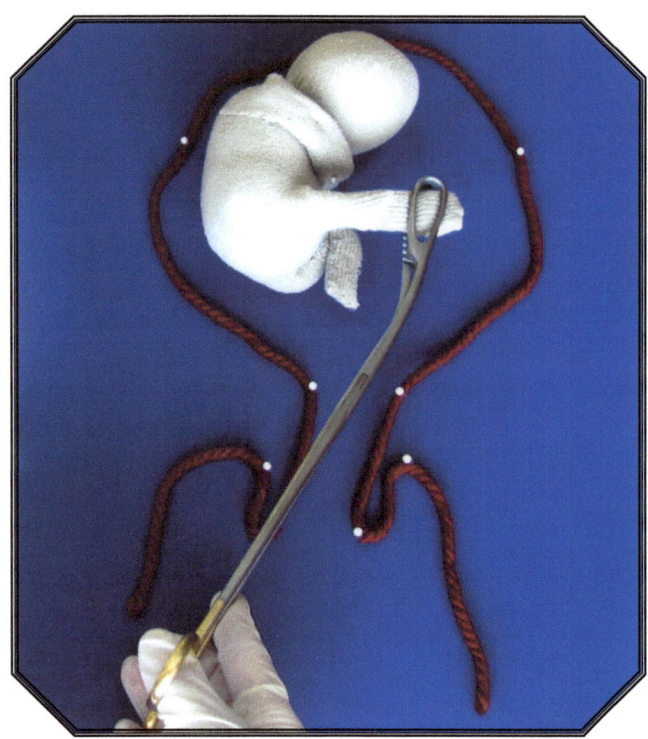

Conversion to a footling breech may require re-grasping.

Approach from patient's right

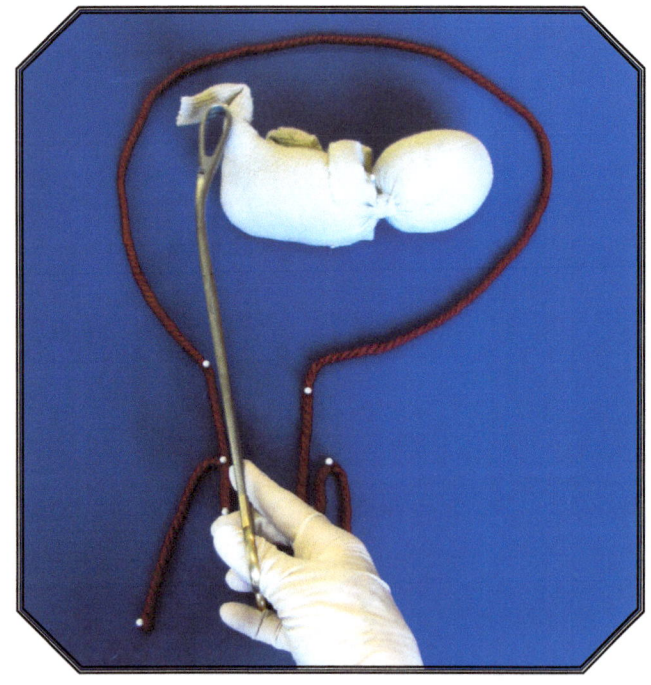

If the lower extremity in a transverse lie is to the patient's right, it may be necessary to approach from the patient's right. Note that, for this approach, the forceps are held on the opposite side from that utilized in an approach from the patient's left or anterior. The forceps are held from the "inside" with the curve pointing to the patient's left. Insert the forceps with the usual technique, and place them to the (patient's) right of the fetus.

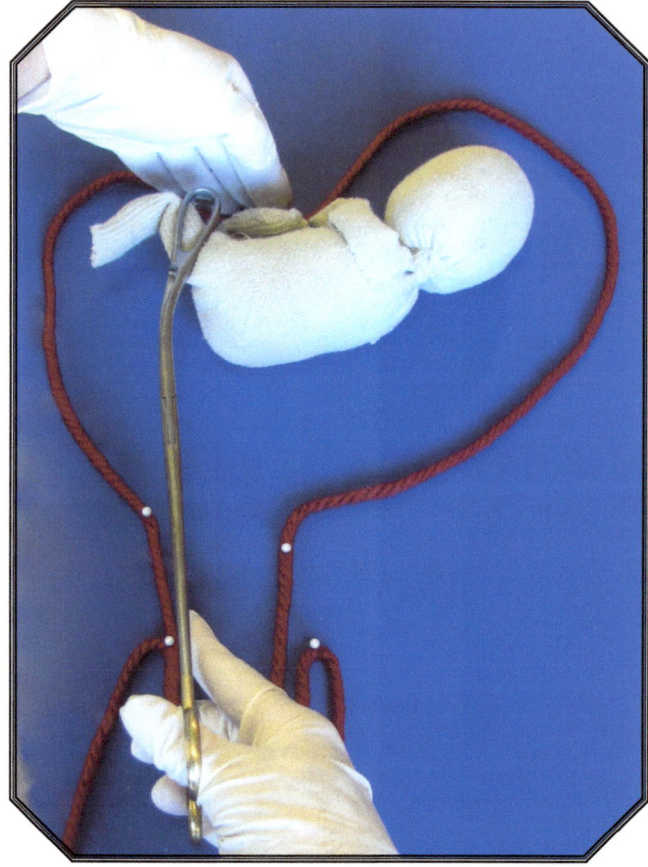

A Hanson maneuver is utilized to facilitate grasping of the lower extremity. Exert traction while turning to bring the part down as described previously.

Since this approach is difficult, an anterior approach to the upper extremity, allowing conversion to a shoulder presentation should be considered.

Conversion to shoulder presentation

When managing transverse lie by converting to shoulder presentation, an anterior approach is best. A lateral approach may be effective if you are converting a vertex presentation or oblique lie.

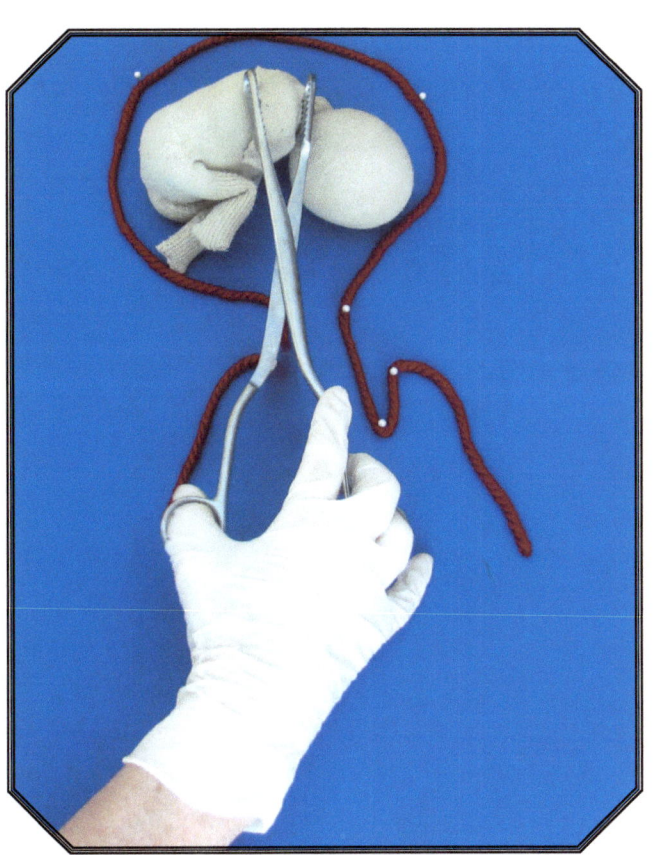

Open, rotating if necessary.

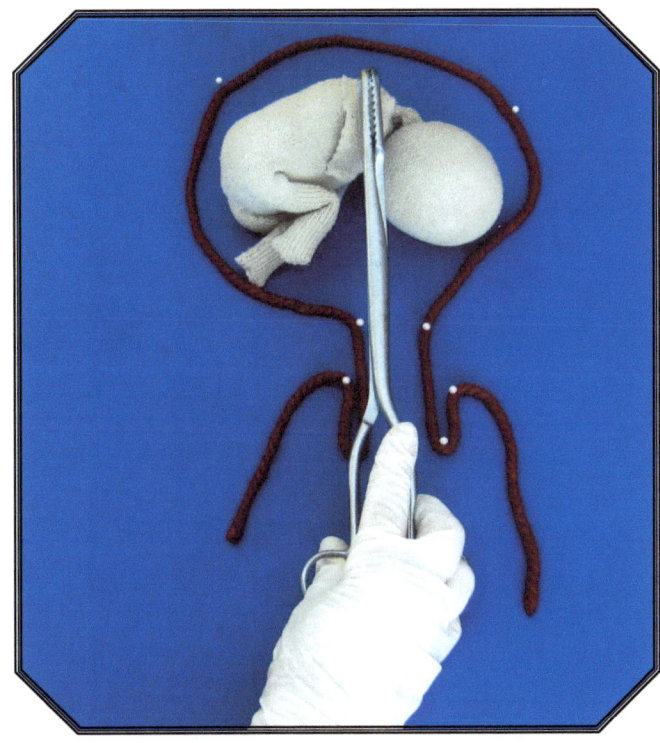

Again, if the uterus is very contracted and it is difficult to position the forceps, maneuvering them to over the upper extremity from either side or below with blades pointing to the patient's right, then turning to point posterior, may be effective.

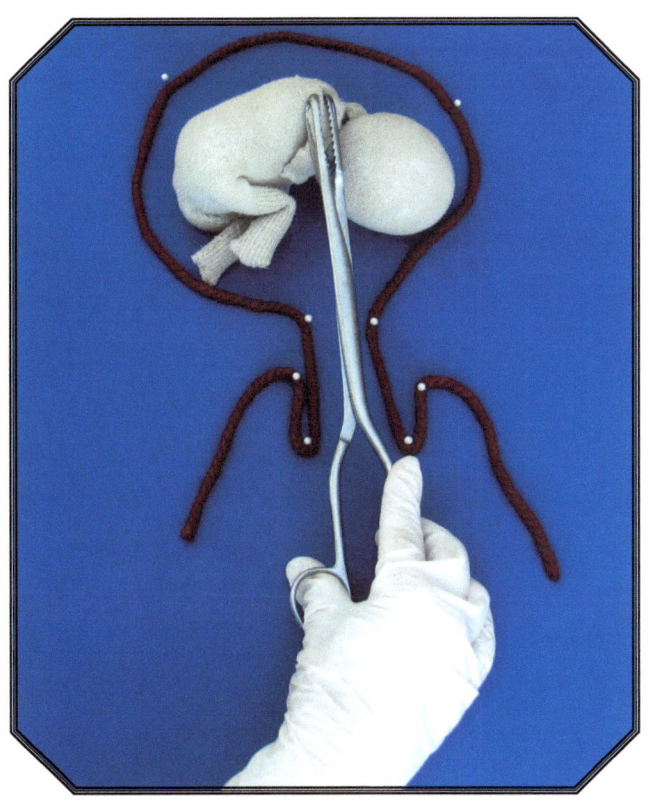

Grasp.

not result in the part descending through the internal os and cervix, I use a "turn-and-pull" technique to extract and separate the grasped part. This technique is more useful in advanced gestations but may be utilized in any situation to avoid exerting excessive force on the cervix during extraction.

This maneuver requires both hands. Continually and repeatedly rotate the forceps 360 degrees clockwise while pulling. You should experience moderate resistance to the turning as well as the pulling.

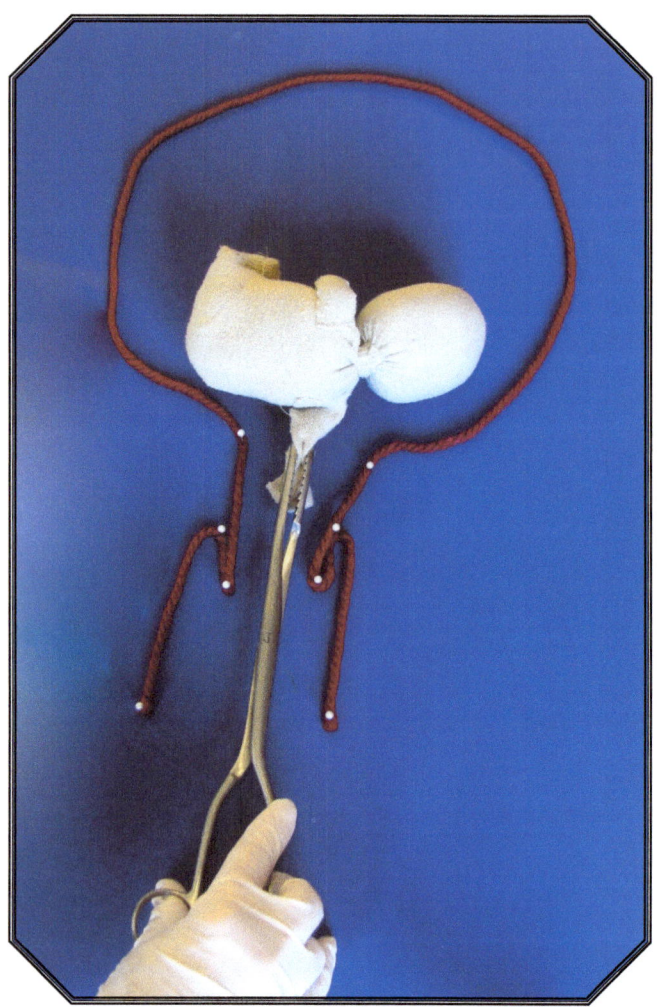

Bring the part down and into the cervix and begin extraction.

Bring the grasped part down through the cervix and as far into the vagina as can be accomplished with moderate traction.

Extraction with "turn-and-pull" technique

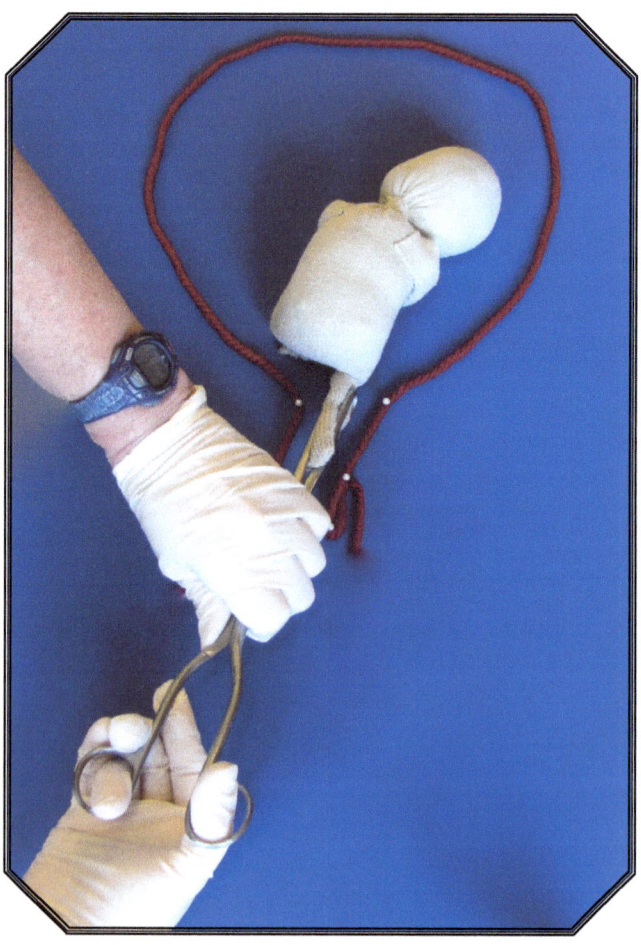

In most situations, a part will separate with gentle to moderate straight outward traction. It is difficult to describe how hard is too hard to pull. Put one arm only into the pull. Again, do not lean backward or brace your foot against the operating table. If moderate traction does

When you need to change the position of your hand, every 180 degrees, hold the shanks at or above the lock firmly together with your left hand.

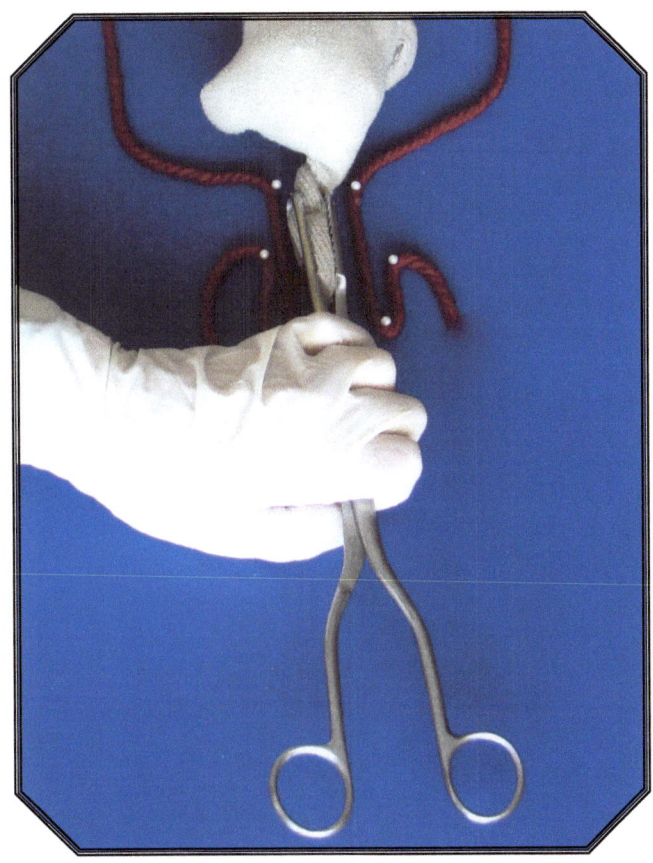

Remove your hand from underneath the forceps.

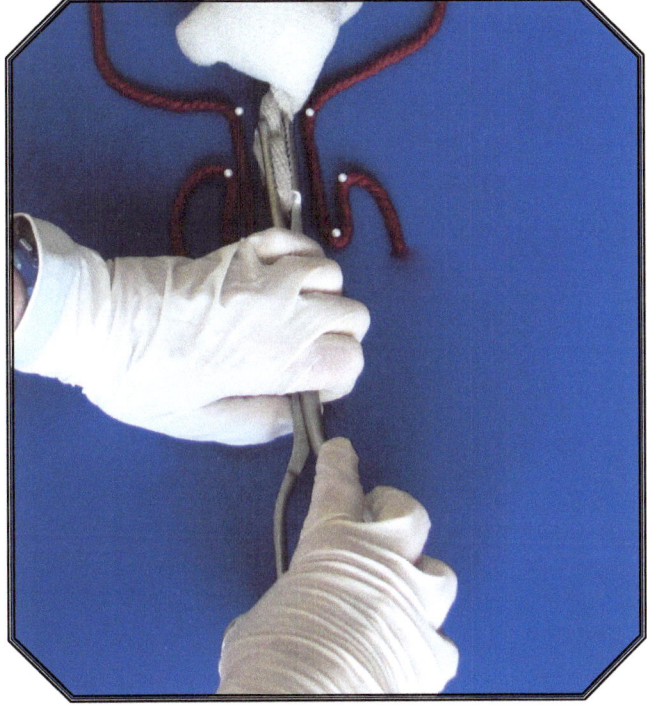

Re-grasp the handles from above, then resume turning and pulling.

If your hands or arms tire, take a break and hold the forceps in place with your left hand. I believe this maneuver exerts more even pressure on the internal os than does a "straight pull" and causes more gradual dilation. It results in "twisting off" the grasped portion.

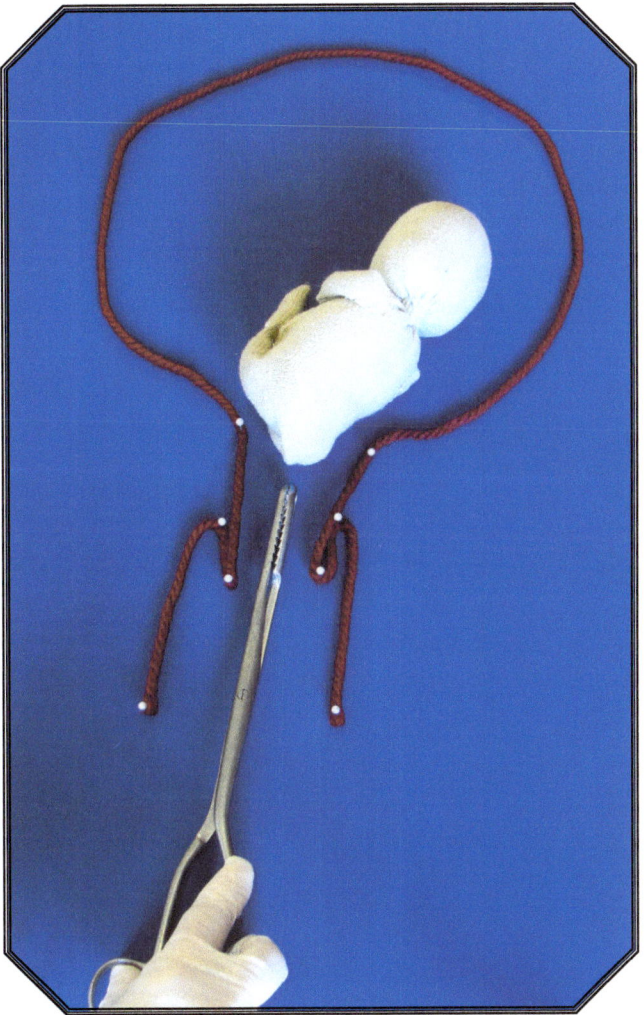

The part that follows is generally firmly fixed into the internal os and can be grasped and removed with the following maneuver.

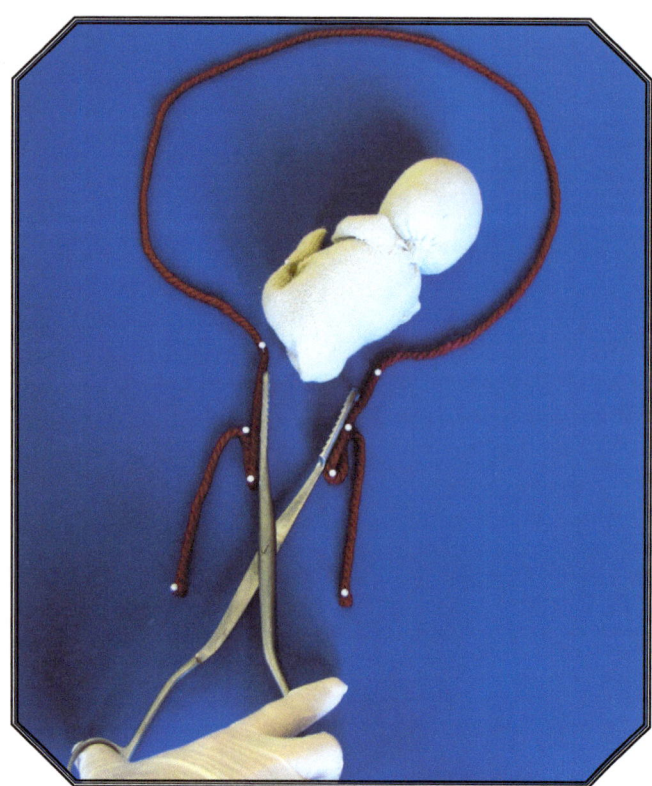

Introduce the forceps into the cervical canal and open them enough to put slight pressure on the sides of the cervical canal.

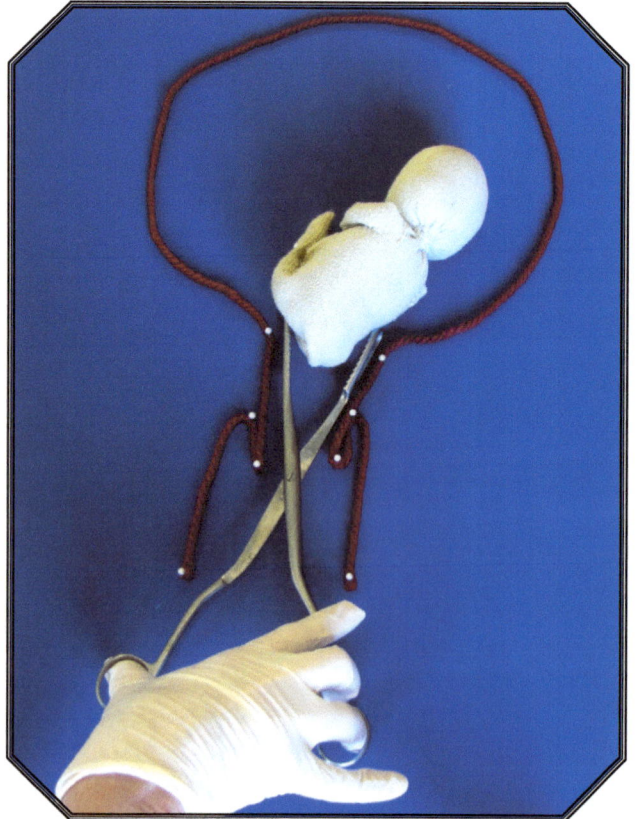

Advance the forceps while you maintain pressure until you encounter the presenting part. Increase the pressure enough to advance the forceps around the presenting part. Continue opening and advancing while you pass the internal os and can visualize the forceps around the presenting part on ultrasound.

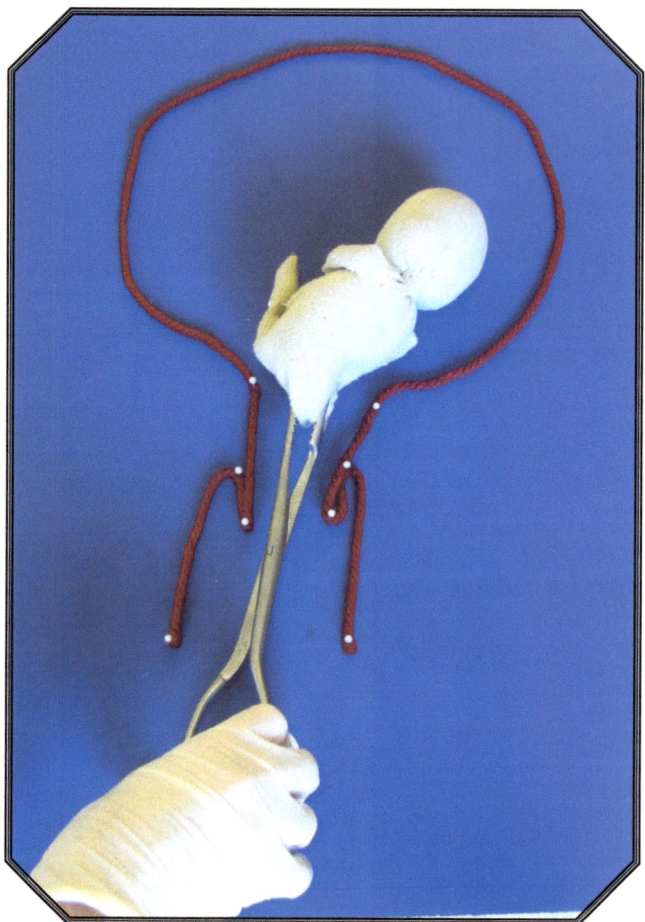

Grasp and extract as described previously.

Extraction of remaining parts

In general, I work in the lower uterine segment to remove the remainder of the small parts and trunk, since this is felt to reduce the risk of perforation. If there is no part immediately accessible at or just above the internal os, visualize one that is higher. It is generally easier to grasp a part, even one on the patient's right, from the patient's left. Since the uterus contracts during the procedure, and the cervix often descends, you may wish to change to a shorter forceps as the case progresses.

For a breech presentation or extraction, when the cervix becomes sufficiently dilated, a standard breech extraction, as described in obstetrical texts, may be initiated.

It is my preference to remove the calvarium at or near the end of the procedure. The cervix is more dilated at this time, and contraction of the uterus results in less space for it to "slip away" (pages 23-24).

Extraction techniques for "entrapped" calvarium

Cervical entrapment

Illustrated below is the extraction of the calvarium of a breech presentation. These techniques are also applicable to a vertex presentation or a detached calvarium. For all presentations, when the plan is to remove the calvarium at or near the end, attempt to leave some tissue attached to the calvarium, and bring the tissue through the cervix to help prevent the calvarium from moving backwards into the fundus during maneuvers. At times, the internal os may contract, causing difficulty inserting the forceps (next page).

Leave any tissue attached to the calvarium protruding through the cervix. If the forceps can be easily inserted, do so with the usual technique. An anterior approach is best for more advanced gestations, or if the calvarium only is remaining.

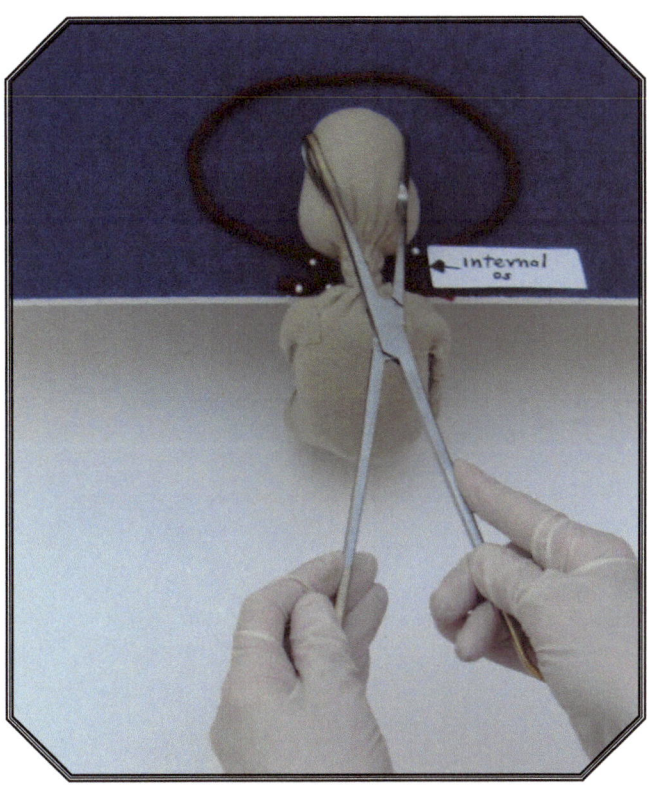

It is oftentimes necessary to use both hands to apply enough force to compress the calvarium in an advanced gestation.

If the internal os has contracted to the point where it is difficult to insert the forceps, utilize the following maneuver.

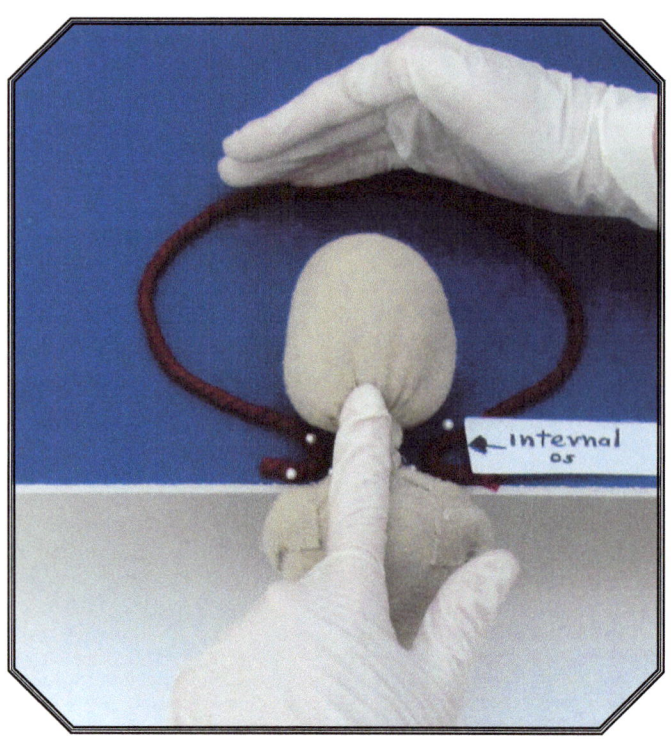

Utilize fundal pressure. Insert your left index finger over the fetal neck. "Wiggle" it forward until the tip has passed the internal os and is over the lower calvarium.

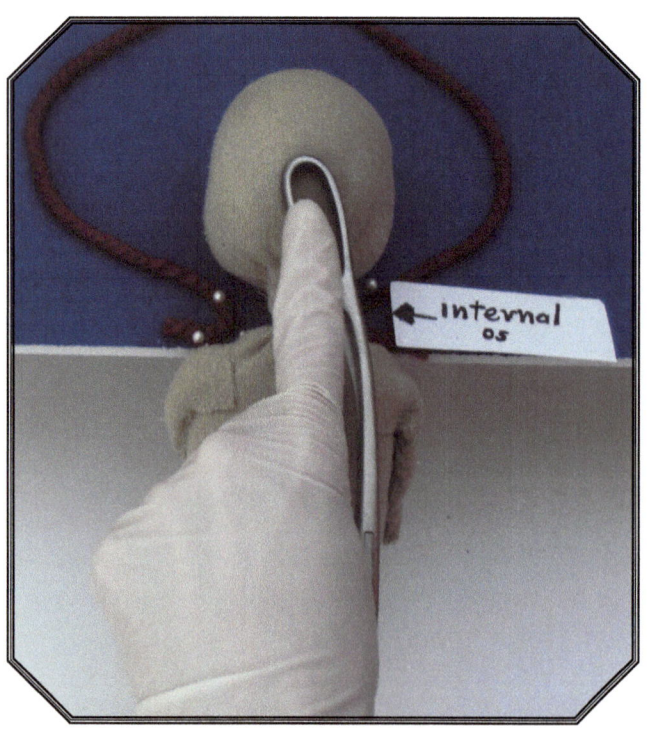

Insert the forceps pointing to the patient's right between the fetal neck and your finger, exerting moderate pressure anteriorly with your finger, and wiggle the blades of the forceps past the os. Remove your finger.

(This maneuver can also be utilized to introduce instruments in a situation where the cervix can be palpated but not visualized.)

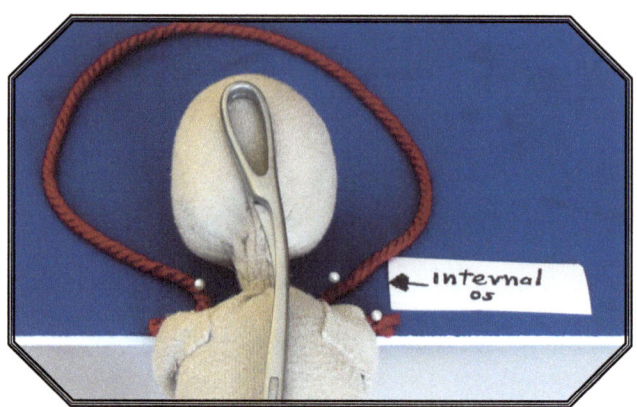

Advance the forceps to slightly past the midpoint of the calvarium.

Proceed with an approach from anterior (page 21).

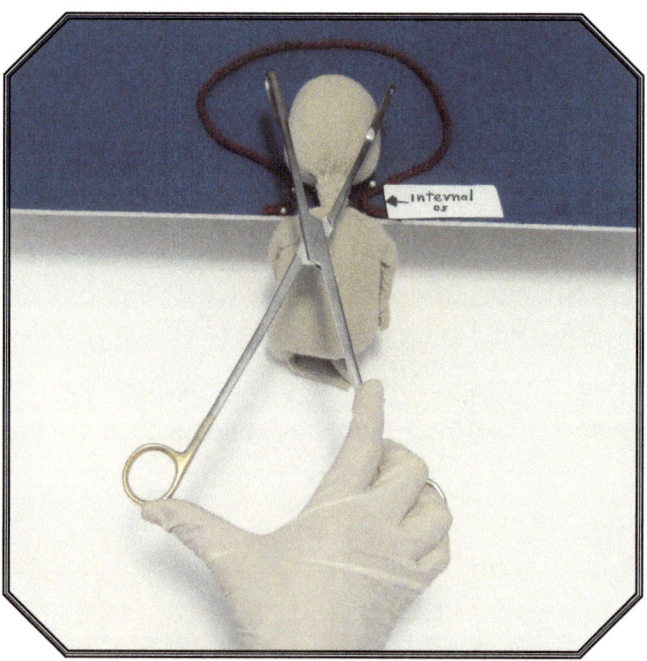

Note this alternate position of the hand, which allows you to open the forceps farther. It may require both hands to fully open the forceps. Advance posterior, grasping the calvarium, and proceed to compression, as previously described.

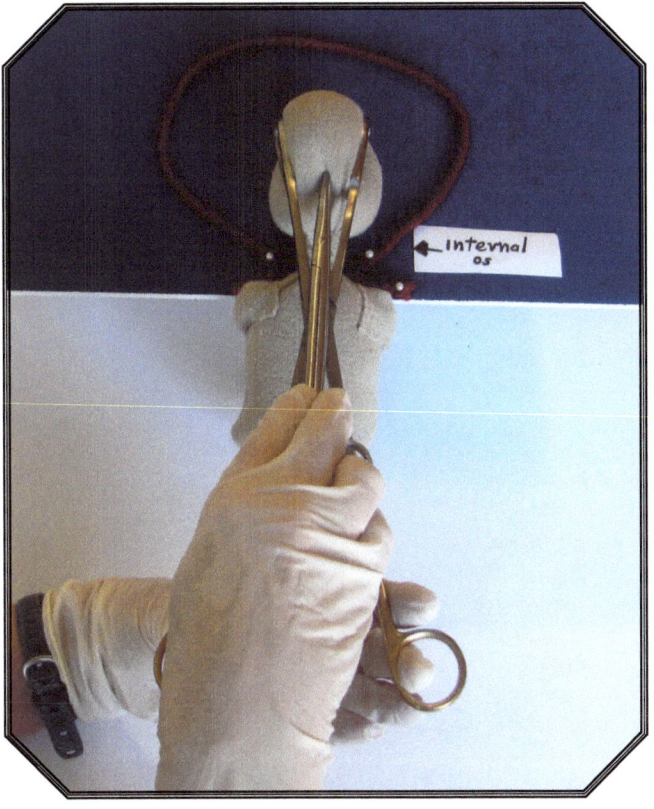

Once decompression has been achieved, rotating continuously through a 360-degree arc while applying traction will facilitate removal.

Cornual entrapment

Removal of the calvarium, or any other part from the cornua at the end of the case, may require some very difficult maneuvering. This problem often occurs when the cervix is retracted anteriorly from a previous Cesarean. Remember that it is perfectly acceptable to allow uterotonic drugs to abate and return the patient to the OR after waiting 2 hrs., at which point the part has generally moved to the lower uterine segment. Very often the internal os has contracted, as has the lower uterine segment. Dilation with large dilators will allow introduction of instruments.

I have, on occasion, found it necessary to pierce the calvarium with a sharp, pointed scissor or pointed scalpel to achieve decompression.

Attempt first to utilize the suction cannula to grasp the part and move it lower in the uterus. If unsuccessful, select a forceps of appropriate length with a pelvic curve. Ultrasonic visualization in the sagittal plane should be utilized for the introduction of instruments.

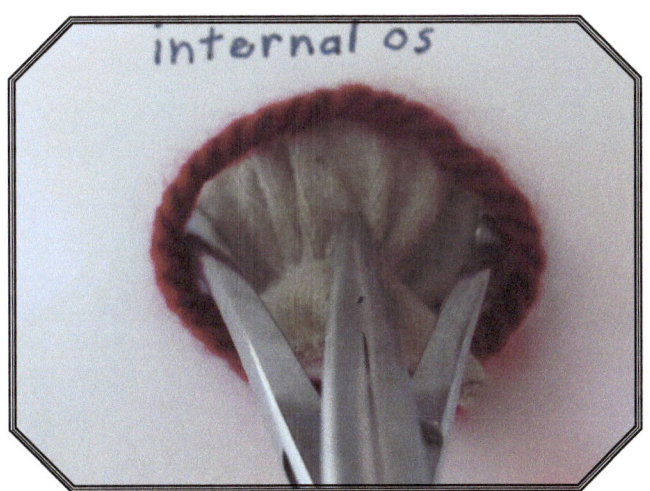

Utilizing slow, steady pressure against the internal os and lower uterine segment, similar to that utilized for dilating a "tough" internal os during a first trimester case, will result in the advancement of the forceps.

If you are using a scalpel, be sure you are able to visualize the blade directly. A small dilator or suction cannula may also be used.

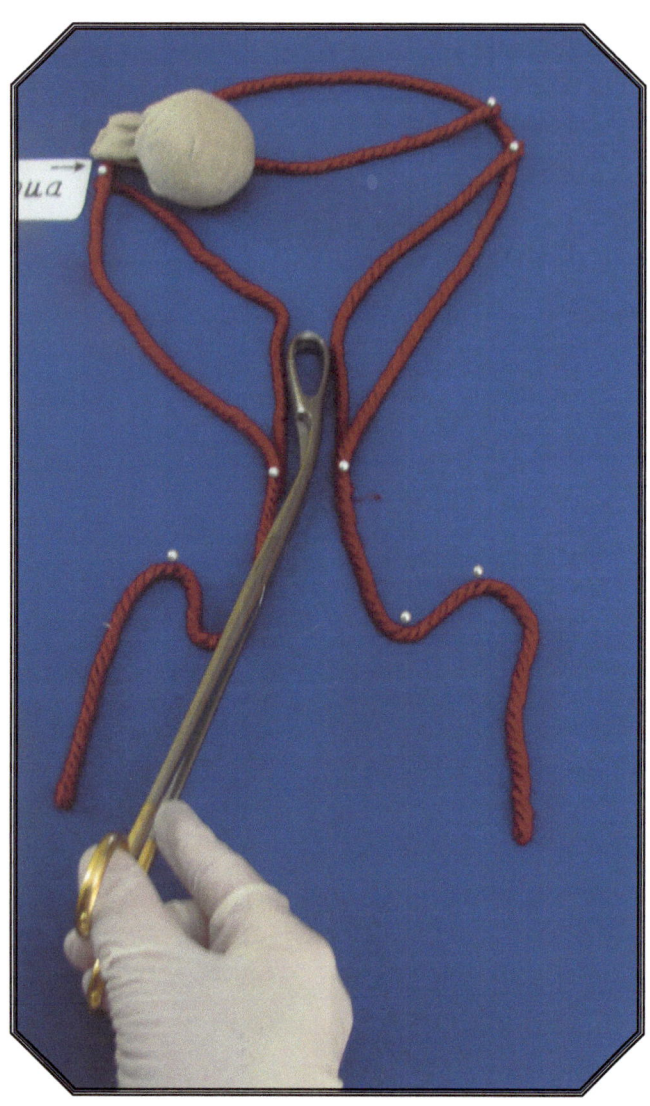

Fundal pressure should be applied while advancing. Because these structures were previously dilated, they have more "give" than you might expect.

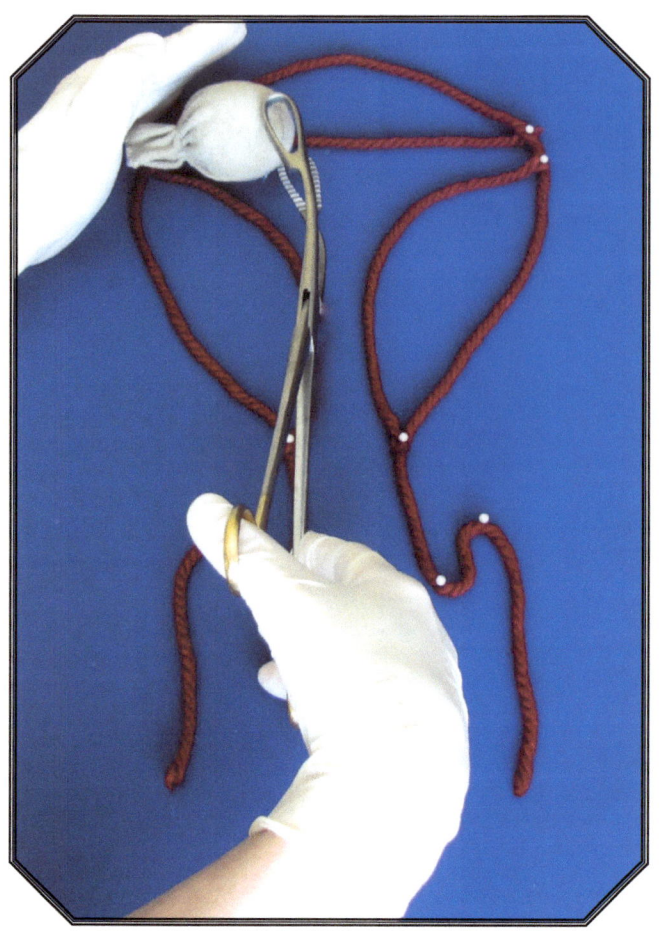

Once you have reached the entrapped part, you will initially be unable to open the forceps. Continue to exert slow, steady pressure to open. This may take several minutes. Utilize your left hand to perform a Hanson maneuver or instruct an assistant to push the part into the forceps blades then close gently.

Even a tenuous grasp may be effective to move the part to a more accessible location. Numerous steps and different approaches may be required to move the part to a location where it can be adequately grasped for extraction.

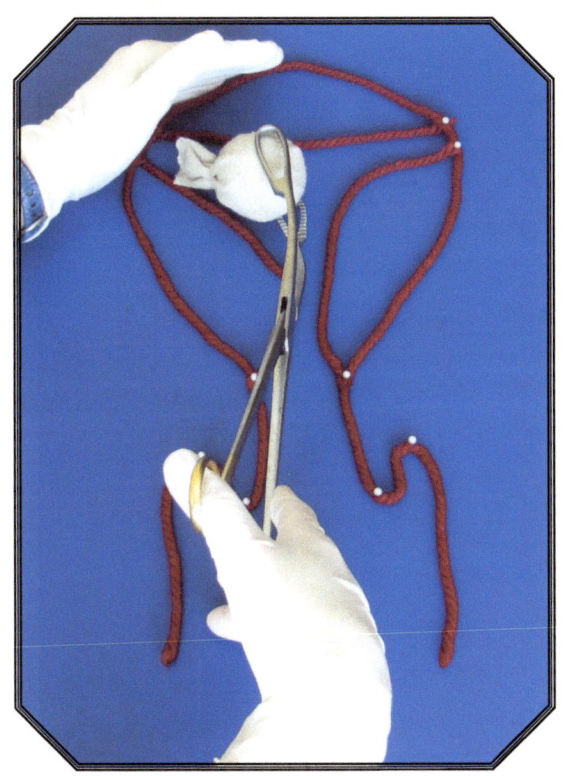

Re-grasp as necessary.

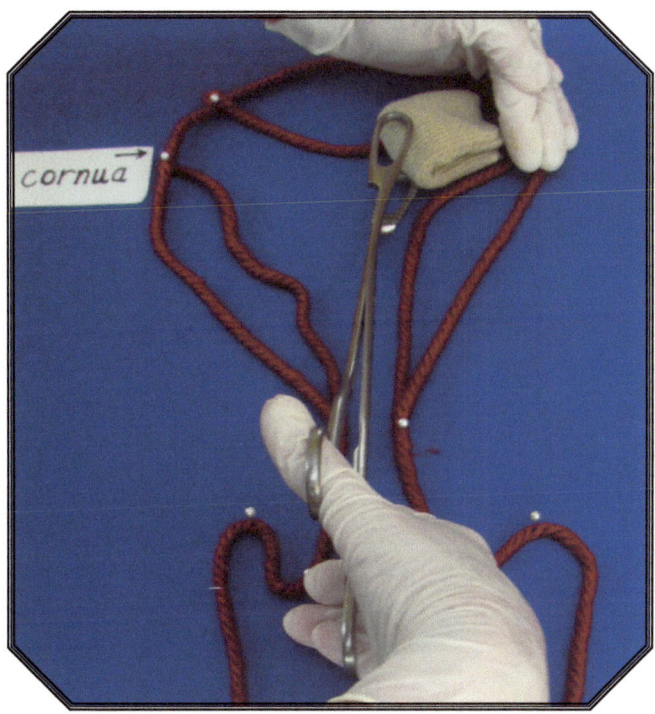

If the part is trapped in the patient's left cornua, an approach from the patient's right will be required. Note the position of the right hand. Change the position of your hand as soon as an approach from anterior is feasible. Proceed as previously described.

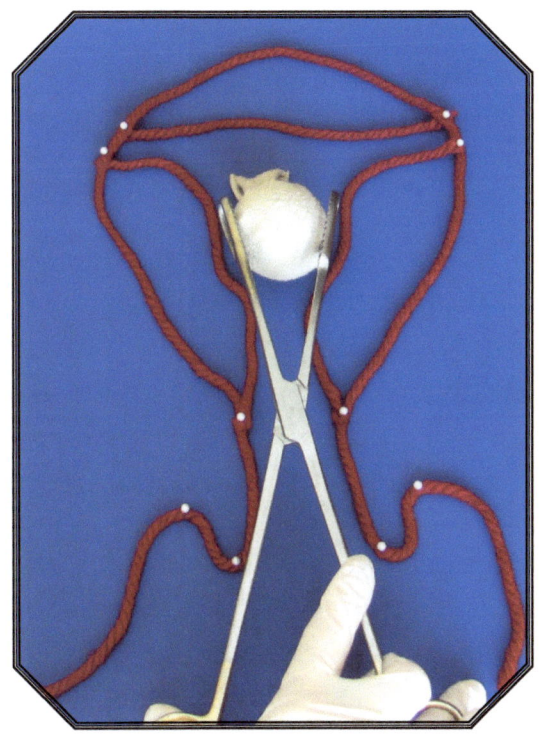

Once the part has been adequately grasped, use the usual turn and pull maneuver to bring the entrapped part through the lower uterine segment and cervix.

Removal of placenta

Removal of the placenta may take place at any time during the case, depending on location. Low-lying placentas may be removed initially. Oftentimes as the uterus contracts, the placenta will separate, and you will grasp it while targeting a fetal part. Bring the placental tissue through the external os, then release and re-grasp above the internal os and pull gently to deliver the entire placenta. It is generally easy to grasp and remove the placenta at the end of the case if it has not previously separated. Suction can be utilized to remove the placenta from the fundus if the uterus has contracted.

Completion of procedure

I generally perform a gentle sharp curettage of the cavity with a large curette to confirm complete evacuation and explore for uterine abnormalities, such as fibroids or septate uterus, which could impede complete emptying, followed by suction curettage. If the case has been performed utilizing transverse ultrasound, the uterus should be scanned in sagittal. If sagittal ultrasound has been utilized, a transverse scan should be performed. All instruments are then removed. Digital examination of the cervical canal is not routine, but should be performed if there is excessive bleeding or any question of cervical injury.

Management of post-op bleeding

I do not routinely give postoperative uterotonic drugs. If the patient is bleeding heavier than normal, methergine 0.2 mg may be injected into the cervix. This may be repeated if necessary. Additional misoprostol, 800-1000 mcg may be given rectally to control bleeding. Oxytocin may be useful in more advanced gestations. Foley catheters with a 30-cc balloon should be available for tamponade or compression of bleeding from the internal os or lower uterine segment.

I previously utilized vasopressin 8 IU in 4 cc lidocaine, injected into the cervix at 3 locations, repeated in several minutes as necessary if the blood pressure remained normal, and found it an excellent treatment for bleeding. If you are utilizing this treatment, be very aware that bleeding will recur when the vasopressin wears off if the underlying cause has not been treated.

It is very important to perform a thorough examination to determine the source of bleeding and address any specific problems. A digital examination of the cervix and lower uterus should be performed, and any lacerations sutured or a balloon applied if suturing is not practical. Introduction of a suction cannula may help to localize bleeding. The patient should be examined ultrasonically for intra-abdominal hematomas, intraperitoneal fluid or other evidence of perforation. If the cavity is enlarged, uterotonic drugs should be given as above and any clots or blood removed with suction. Patients should be observed carefully for at least 2 hrs.

Transfer to post-op

RhoGAM should be given if indicated, anesthesia terminated and the patient prepared for transfer to the post-anesthesia care area. A skilled RN capable of managing common problems and recognizing complications should observe routine patients for at least one hour prior to discharge. Standard discharge instructions should be given.

References:

(1) Hammond, C. MD, & Chasen, S. MD. (2009). Dilation and evacuation. In Maureen Paul, MD, MPH; E. Steve Lichtenberg, MD, MPH; Lynn Borgatta, MD, MPH; David A. Grimes, MD; Phillip G. Stubblefield, MD; Mitchell D. Creinin, MD (Eds.), *Management of Unintended and Abnormal Pregnancy* (pp. 157-177). Wiley-Blackwell, UK

(2) Library of Congress, Pub.L. 108-105, 117 Stat. 1201, 18 U.S.C. 1531

www.ingramcontent.com/pod-product-compliance
Lightning Source LLC
Chambersburg PA
CBHW040412220526
45473CB00004B/1212